Key Stage 4 Revision

On Course

GCSE Science
Single Award and Double Award
. . . .
TEACHER'S BOOK

ASSESSMENT and
QUALIFICATIONS
ALLIANCE

NORTHERN EXAMINATIONS
and ASSESSMENT BOARD

Stanley Thornes (Publishers) Ltd

First published in 1998 by:
Stanley Thornes (Publishers) Ltd
Ellenborough House
Wellington Street
Cheltenham GL50 1YW
England

98 99 00 01 / 10 9 8 7 6 5 4 3 2 1

A catalogue record of this book is available from the British Library.

ISBN 0-7487-3670-0

Typeset by Mathematical Composition Setters, Salisbury, Wiltshire
Printed and bound in Great Britain by Ashford Colour Press, Gosport, Hampshire
Artwork by Peters & Zabransky

Contents

Science: On Course for GCSE NEAB Edition Teacher's Book © Stanley Thornes (Publishers) Limited 1998

Double Award   **Single Award**

**Answers to Revision Notes
and Summary Questions**

Science: On Course for GCSE NEAB Edition Teacher's Book © Stanley Thornes (Publishers) Limited 1998

Introduction

Science: On Course for GCSE (NEAB Edition) has been written by examiners and endorsed by NEAB for their Modular Science syllabuses (Double Award 1206 or Single Award 1208) and their Co-ordinated Science syllabuses (Double Award 1201 and Single Award 1203).

It is designed to help students to prepare more effectively and thoroughly for their examinations by giving a summary of essential knowledge and plenty of practice with assessment. Students benefit from having their own syllabus revision resource. They can use it over the two years of their course or as a ready-made revision aid prior to their module tests and/or examinations.

The practice questions reflect those set by NEAB from 1998 onwards, for both Modular and Co-ordinated syllabuses. They can be used for each topic when the topic is completed, i.e. throughout Years 10 and 11, or they can be 'saved-up' and used for consolidation and practice prior to module tests and examinations as part of a revision programme.

Science: On Course is efficient and easy to use from the teacher's perspective too, enabling marking, monitoring and diagnosing weaknesses with individual students and whole teaching groups.

The Revision Notes are headed with a set of Key Words to reflect the importance of understanding and using correct scientific terms. Students should complete the cloze passages for each topic shortly after they have completed the topic in class, or at the end of the course. The Key Words are defined in the Glossary and Answers are given in the student's book to promote confidence and success, so no student is unsupported.

The Revision Notes could be completed at home as a homework activity or in class. There are a number of ways of marking and giving feedback. You could mark the Revision Notes in the traditional way and give a total mark. Alternatively, students could mark their own Revision Notes using the set of answers in this book (note they can be detached from the mark schemes for GCSE questions if required).

Another method of marking and giving feedback involves using the completed cloze passages provided in this book. You can produce an OHT Sheet of these pages and use them as overlays for rapid marking. Alternatively, the method recommended by the authors is to use the OHT Sheets with the class; reinforcing important teaching points and getting the students to amend their answers as necessary.

Practice with mock GCSE questions is becoming increasingly important, especially those questions designed to assist development of continuous and extended writing answers. Mark Schemes including examiner's advice are also provided in a form for students to use.

This Teacher's Book includes two Record Sheets; one for recording an individual student's performance and the other for keeping class records. They will assist in identifying areas of individual and group weaknesses. Teachers may convert the Record Sheets into a computer spreadsheet.

We believe that **Science: On Course** will, by active and planned revision, raise the levels of motivation, success and achievement of students in GCSE examinations. We hope you will find it a useful and flexible way to help your students improve their results at GCSE.

Bob McDuell
Keith Hirst
Graham Booth

Acknowledgements

The authors and publisher would like to thank the following for supplying photographs:

Britstock-IFA: p. 49 (TPL); p. 52 (Eric Bach)
Brookes & Vernons PR: p. 83
Dr Olaf Linden/ICCE: p. 43
Francis Gohier/Ardea: p. 32
GeoScience Features: pp. 40, 56, 96
Heather Angel: p. 26
Holt Studios International: p. 54 (Andy Burridge)
Martyn Chillmaid: pp. 38, 65, 74
Rotary Burnand: p. 76
Science Photo Library: p. 8 (Biophoto Associates); p. 16 (Dick Luria); p. 46 (Peter Menzel)

Science: On Course for GCSE

Topic: _____ Class/Form/Set: _____

Teacher: _____

STUDENT

QUESTION

| 1 |
| 2 |
| 3 |
| 4 |
| 5 |
| 6 |
| 7 |
| 8 |
| 9 |
| 10 |
| 11 |
| 12 |
| 13 |
| 14 |
| 15 |
| 16 |
| 17 |
| 18 |
| 19 |
| 20 |
| 21 |
| 22 |
| 23 |
| 24 |
| 25 |
| 26 |
| 27 |
| 28 |
| 29 |
| 30 |
| 31 |
| 32 |
| 33 |
| 34 |
| 35 |
| 36 |
| 37 |
| 38 |
| 39 |
| 40 |
| 41 |
| 42 |
| 43 |
| 44 |
| 45 |
| 46 |
| 47 |
| 48 |
| 49 |
| 50 |
| 51 |
| 52 |
| 53 |
| 54 |
| 55 |
| 56 |
| 57 |
| 58 |
| 59 |
| 60 |
| 61 |
| 62 |
| 63 |
| 64 |
| 65 |
| 66 |
| 67 |
| 68 |
| 69 |
| 70 |
| 71 |
| 72 |
| 73 |
| 74 |
| 75 |

Science: On Course for GCSE

TEACHER'S RECORD SHEET

Student: _____

Class/Form/Set: _____

Teacher: _____

TOPIC	Life processes …	Nutrition	Breathing …	Circulation …	Control …	Homeostasis	Photosynthesis …	Water relations	The environment	Energy flow …	Variation	Inheritance …	Metals …	Acids, bases …	Rocks in the Earth	Chemicals from oil	The earth …	Rates …	Energy changes …	Chemicals from air	Atomic structure …	The Periodic Table	Transferring energy	Generating…electricity	Current, charge …	Using electricity	Magnetism …	Force and motion	Forces … effects	The Earth …	Wave properties	Electromagnetic …	Radioactivity

QUESTION 1–75

Life processes

Do part of this topic for Single Science and all of it for Double Science.
D = for Double Science only   H = for Higher Tier only

 ## Revision notes

Life processes

🔑	excretion	growth	movement	nutrition
	reproduction	respiration	sensitivity	

The chick has the following life processes in common with all living organisms:

- Obtaining food by eating plants or by eating other animals is ⬛1 _____ .
  nutrition

- Releasing energy from food is ⬛2 _____ .
  respiration

- Releasing waste products is ⬛3 _____ .
  excretion

- Producing offspring is ⬛4 _____ .
  reproduction

- Development from young to adult is ⬛5 _____ .
  growth

- Reacting to the surroundings is ⬛6 _____ .
  sensitivity

- Changing position is ⬛7 _____ .
  movement

Structure of cells

🔑	cell membrane	cell wall	cytoplasm
	genes	nucleus	protein coat

The chick is made up of cells. Cells have many common features.

Use key words to label these drawings of an animal cell, the bacterium and the virus.

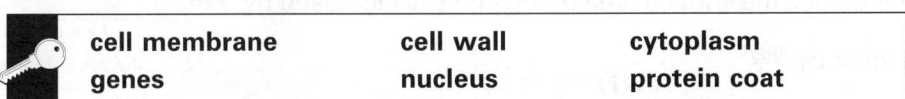

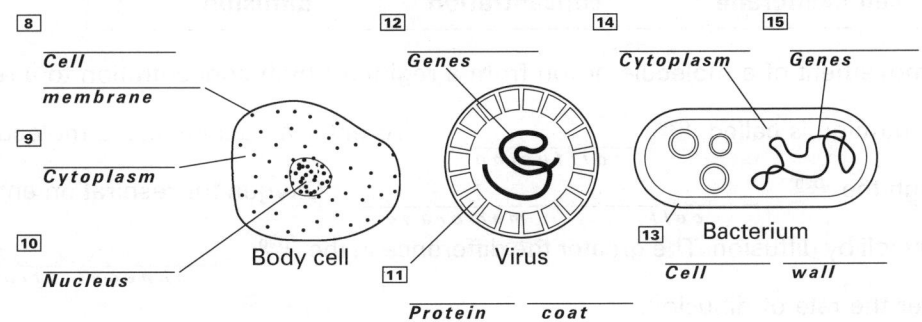

Science: On Course for GCSE NEAB Edition Teacher's Book © Stanley Thornes (Publishers) Limited 1998

Functions of cell parts

🔑	cell membrane	cell wall	cytoplasm	nucleus

Complete this table showing the jobs of the parts of cells:

job	part of cell
controls the activities of the cell	[16] _____ *nucleus*
where most of the chemical reactions occur	[17] _____ *cytoplasm*
controls the passage of substances in and out of the cell	[18] _____ _____ *cell* *membrane*
strengthens bacterial cells	[19] _____ _____ *cell* *wall*

Groups of cells

🔑	contract	glandular	organ	organ system	tissue
	alveoli	surface area	villi		

A group of cells with a similar structure and a particular job is called a [20] _____ . *tissue*

The job of muscular tissue is to [21] _____ . The job of [22] _____ tissue *contract* *glandular*

is to produce useful substances. Groups of tissues are called an [23] _____ . Different *organ*

organs working together form an [24] _____ _____ . *organ* *system*

Some organ systems are specialised for exchanging by having an increased [H25] _____ *surface*

_____ . In humans the surface area of the lungs is increased by [H26] _____ , and *area* *alveoli*

that of the small intestine by [H27] _____ . *villi*

Exchanging materials

🔑	cell membrane	concentration	diffusion

The movement of a molecule or ion from a region of high concentration to a region of lower

concentration is called [D28] _____ . To enter an animal cell a molecule or ion must pass *diffusion*

through the [D29] _____ _____ . Oxygen for respiration enters the body through *cell* *membrane*

the alveoli by diffusion. The greater the difference in the [D30] _____ of oxygen, the *concentration*

greater the rate of diffusion.

Nutrition

Do all of this topic for Single Science and Double Science.
H = for Higher Tier only

 ## Revision notes

Food

carbohydrate	cell membrane	energy	fat	growth	protein

The cheeseburger contains meat, cheese and bread (which is made from cereals).

Cereals, fruits and root vegetables are rich in $\boxed{1}$ _____ which is needed to
                                  carbohydrate

provide $\boxed{2}$ _____ . Meat, fish, eggs and pulses are rich in $\boxed{3}$ _____ which is
           energy   *protein*

needed for $\boxed{4}$ _____ and for replacing cells. Milk, cheese, butter and margarine are rich
             growth

in $\boxed{5}$ _____ which is needed to provide $\boxed{6}$ _____ and for making
      fat   *energy*

$\boxed{7}$ _____ _____ .
    cell   *membranes*

Structure of the digestive system

anus	gullet	large intestine	liver	pancreas
small intestine	stomach			
gall bladder				

Label the parts of the digestive system.

$\boxed{H15}$
Gall bladder

$\boxed{8}$
Gullet

$\boxed{9}$
Liver

$\boxed{10}$
Stomach

$\boxed{11}$
Pancreas

$\boxed{12}$
Small intestine

$\boxed{13}$
Large intestine

$\boxed{14}$
Anus

Science: On Course for GCSE NEAB Edition Teacher's Book © Stanley Thornes (Publishers) Limited 1998

Digestion

bloodstream	enzyme	insoluble	muscular	soluble

We need the digestive system to break down large [16] _____ food molecules into
insoluble
smaller [17] _____ molecules that can be absorbed into the [18] _____ .
soluble   *bloodstream*
This breakdown of food is speeded up by proteins called [19] _____ .
enzymes
Food is moved through the digestive system by the contraction of [20] _____ tissue.
muscular

Enzymes

amino acid	bacteria	bile	carbohydrase	fatty acid
glandular	hydrochloric acid	lipase	pancreas	protease
salivary gland	soluble	stomach	sugar	
acidic	*alkaline*	*emulsifies*	*gall bladder*	*lipase*
neutralise	*small intestine*	*surface area*		

Enzymes are produced by [21] _____ tissue.
glandular
The enzyme that breaks down starch into [22] _____ is called [23] _____ .
sugars   *carbohydrase*
The enzyme that breaks down proteins into [24] _____ _____ is called
amino   *acids*
[25] _____ .
protease
The enzyme that breaks down fat into [26] _____ _____ and glycerol is
fatty   *acids*
called [27] _____ . Complete this table:
lipase

enzyme	produced in		
amylase	[28] _____ _____ , pancreas and small intestine		
	salivary *gland*		
proteases	[29] _____ , pancreas and small intestine		
	stomach		
lipase	[30] _____ and small intestine		
	pancreas		

In addition to enzymes, the stomach produces [31] _____ _____ to
hydrochloric   *acid*
kill [32] _____ in the food.
bacteria
The liver produces a green liquid called [H33] _____ . This is stored in the
bile
[H34] _____ _____ until needed. Following a meal, bile passes into the
gall   *bladder*
[H35] _____ _____ and mixes with the food. As this food has just left
small   *intestine*
the stomach it is [H36] _____ . Bile is [H37] _____ in order to
acidic   *alkaline*
[H38] _____ the food so that all the enzymes in the small intestine can work more
neutralise
effectively.

Most fats melt in the stomach to form large droplets. Bile [H39] _____emulsifies_____ these large droplets into smaller ones. This process means that fats have a larger [H40] _____surface_____ _____area_____ for the enzyme [H41] _____lipase_____ to act on.

Absorption

amino acid	anus		faeces	glycerol	large intestine
sugar	water
surface area	*villus (plural villi)*

Digestion of food is completed in the small intestine.

Complete this table:

type of food	soluble end products of digestion
carbohydrates	[42] _____sugars_____
proteins	[43] _____amino_____ _____acids_____
fats	fatty acids and [44] _____glycerol_____

The inner surface of the small intestine is folded and has finger-like projections called [45] _____villi_____ which increase the [46] _____surface_____ _____area_____ for the absorption of soluble food.

Indigestible food passes from the small intestine into the [47] _____large_____ _____intestine_____ where most of the [48] _____water_____ is absorbed from it, forming [49] _____faeces_____ . These are passed out of the body via the [50] _____anus_____ .

Science: On Course for GCSE NEAB Edition Teacher's Book © Stanley Thornes (Publishers) Limited 1998

Breathing and respiration

Do all of this topic for Double Science only.
H = for Higher Tier only

 Revision notes

Structure of the thorax

| alveolus (plural alveoli) | bronchiole | bronchus | diaphragm |
| lung | rib | rib muscle | trachea |

This diagram shows the organs inside the human thorax. Use key words to label these organs.

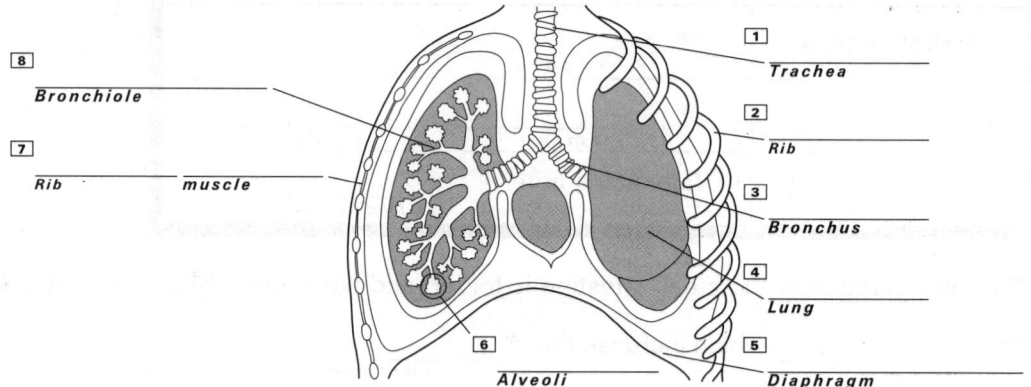

8
Bronchiole

7
Rib   muscle

1
Trachea

2
Rib

3
Bronchus

4
Lung

5
Diaphragm

6
Alveoli

Breathing and respiration

active transport	aerobic	anaerobic	carbon dioxide	energy
lactic acid	muscle	oxygen		
capillary	*fatigue*	*lactic acid*	*muscle*	*oxidised*
oxygen debt	*pressure*	*surface area*	*volume*	*water*

The runner needs to breathe to take in the oxygen to release the energy he needs to run.

To breathe in, the ^{H9} _____ between the ribs contract pulling the ribs upwards. At the
muscles

same time the diaphragm muscles contract pulling the diaphragm ^{H10} ~~upwards~~/downwards.

These movements increase the ^{H11} _____ of the thorax, causing a decrease in
volume

^{H12} _____ . Air then moves into the lungs. In the alveoli, ¹³ _____ diffuses from
pressure   *oxygen*

the air into the blood and ¹⁴ _____ _____ diffuses from the blood into the air.
carbon   dioxide

Science: On Course for GCSE NEAB Edition Teacher's Book © Stanley Thornes (Publishers) Limited 1998

Respiration is the process by which [15] _____*energy*_____ is released from food.

[16] _____*aerobic*_____ respiration uses oxygen, but [17] _____*anaerobic*_____ respiration does not use oxygen.

Complete the word equation for aerobic respiration:

Glucose + oxygen → [18] _____*carbon*_____ _____*dioxide*_____ + water + energy.

Complete the word equation for anaerobic respiration in muscle cells:

Glucose → [19] _____*lactic*_____ _____*acid*_____ + energy.

The energy released during respiration is used to make [20] *large/small* molecules, to enable

[21] _____*muscles*_____ to contract and to move materials across boundaries by [22] _____*active*_____

_____*transport*_____.

Aerobic respiration releases far [H23] *more/less* energy than anaerobic respiration.

If muscles are used vigorously for long periods they begin to suffer from muscle

[H24] _____*fatigue*_____.

If insufficient oxygen reaches a muscle, anaerobic respiration occurs in which glucose is not

[H25] _____*oxidised*_____, resulting in an [H26] _____*oxygen*_____ _____*debt*_____. To repay this debt,

[H27] _____*lactic*_____ _____*acid*_____ is oxidised to carbon dioxide and [H28] _____*water*_____.

Alveoli are specialised for gaseous exchange. They are moist, thin-walled, folded to increase their

[H29] _____*surface*_____ _____*area*_____ and are well supplied with [H30] _____*capillaries*_____.

Science: On Course for GCSE NEAB Edition Teacher's Book © Stanley Thornes (Publishers) Limited 1998

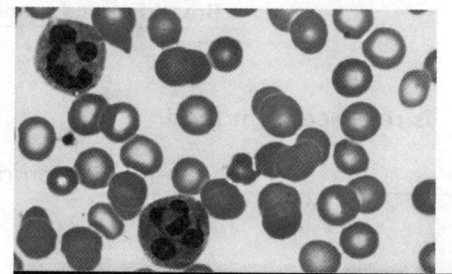

Circulation and defence

Do part of this topic for Single Science and all of it for Double Science.
D = for Double Science only   H = for Higher Tier only

📝 Revision notes

Transport by the blood

haemoglobin	hormone	lung	nucleus	organ
oxygen	plasma	platelet	red cell	small intestine
urea	white cell			

Blood consists of red cells, white cells and platelets suspended in a liquid called [1] _____ .
                                                                                    plasma

[2] _____ _____ are biconcave in shape and contain a red pigment called
      red   *cells*

[3] _____ . They do not possess a [4] _____ . Their job is to transport
   haemoglobin   *nucleus*

[5] _____ from the [6] _____ to the [7] _____ .
   oxygen   *lungs*   *organs*

[8] _____ _____ have a cell membrane, nucleus and cytoplasm.
   white   *cells*

Blood [9] _____ are cell fragments.
         platelets

Complete this table for transport of substances by plasma:

substance	transported from		transported to
carbon dioxide	[10] _____ *organs*		[11] _____ *lungs*
soluble food	[12] _____ _____ *small* *intestine*		liver
[13] _____ *urea*	liver		kidney
[14] _____ *hormones*	glands		organs

What causes diseases

bacteria	infection	toxin	unhygienic	virus

Diseases can be caused when microbes enter the body. Microbes which have a cell wall, membrane

and cytoplasm, but whose genes are not in a distinct nucleus, are called [15] _____ .
                                                                              bacteria

Science: On Course for GCSE NEAB Edition Teacher's Book © Stanley Thornes (Publishers) Limited 1998

Microbes which can only reproduce inside living cells are called [16] _____ .
v i r u s e s

Disease is more likely to occur if we come into contact with a person already carrying an

[17] _____ or if we live in [18] _____ conditions. Microbes may
i n f e c t i o n   *u n h y g i e n i c*

reproduce rapidly inside the body and produce poisons called [19] _____ which make us
t o x i n s

feel ill.

How the body protects us against disease

antibody	antitoxin	clot	hydrochloric acid
immune	ingest	mucus	

White blood cells protect the body in three ways. Some types can [20] _____ microbes.
i n g e s t

Others produce [21] _____ to counteract the poisons produced by microbes.
a n t i t o x i n s

Others produce [22] _____ which kill microbes.
a n t i b o d i e s

Once they have produced antibodies against a particular microbe, white cells can quickly produce

them again so that the body is [23] _____ to that particular disease.
i m m u n e

Platelets protect the body by helping to form a [24] _____ at the site of a wound to prevent
c l o t

the entry of microbes. The skin acts as a barrier against microbes.

[25] _____ _____ produced in the stomach kills microbes present in
h y d r o c h l o r i c   *a c i d*

food.

Microbes in the air are trapped by [26] _____ produced by the breathing passages.
m u c u s

Structure of the heart

artery	atrium (plural atria)	backflow	contract
muscle	valve	vein	ventricle

Label this diagram of the human heart:

Science: On Course for GCSE NEAB Edition Teacher's Book © Stanley Thornes (Publishers) Limited 1998

Checking the pulse is an important medical test. Each pulse is the result of one heart beat.

The heart consists mainly of [D32] _____ tissue which [D33] _____ to
                                    muscle   *contracts*

force blood around the body. It contains valves to prevent the [D34] _____ of blood.
                                                                        backflow

Blood enters the heart via the [D35] _____ . These contract to force blood into the
                                          atria

[D36] _____ which contract to force blood out of the heart.
          ventricles

Circulation

| 🔑 | artery | capillary | carbon dioxide | elastic | muscle |
| | oxygen | tissue fluid | two circulations | valve | vein |

Label the types of blood vessel on this diagram of the circulatory system:

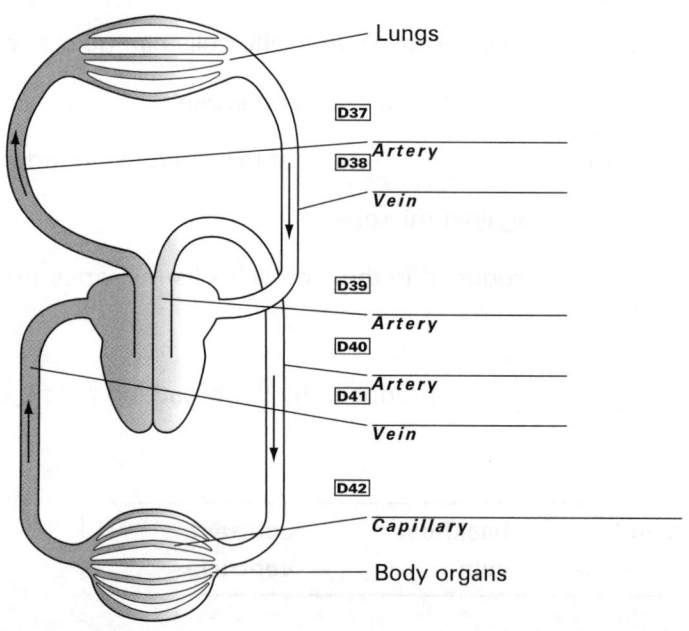

Lungs

[D37]
[D38] *Artery*
Vein
[D39]
Artery
[D40]
Artery
[D41]
Vein
[D42]
Capillary
Body organs

Arteries have thick walls containing [D43] _____ tissue so that they can control the supply of
                                              muscle

blood to the organs. They also have [D44] _____ tissue so that they can expand when the
                                              elastic

heart beats. They contain blood at [D45] *high/low* pressure. Veins have thinner walls and, unlike arteries,

contain [D46] _____ . Capillaries have walls one cell thick to allow [D47] _____
              valves   *tissue*

_____ to flow out carrying food and oxygen to the tissues. All arteries, except the artery
    fluid

carrying blood to the lungs, carry blood that is rich in [D48] _____ . All veins, except the vein
                                                                  oxygen

carrying blood from the lungs to the heart, carry blood that is rich in [D49] _____
                                                                                  carbon

_____ .
    dioxide

Science: On Course for GCSE NEAB Edition Teacher's Book © Stanley Thornes (Publishers) Limited 1998

As blood has to go through the heart twice on one complete journey round the system, the circulation system is said to show D50 _____ _____ .

two *circulations*

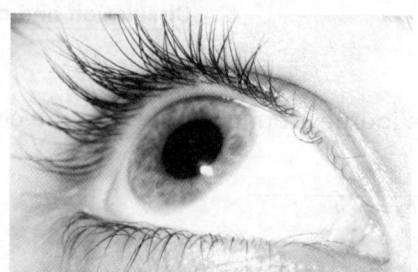

Control and co-ordination

Do all of this topic for Single Science and Double Science.
H = for Higher Tier only

 ## Revision notes

Receptors

ciliary muscle	cornea	ear	eye	iris
lens	nose	optic nerve	pupil	retina
sclera	skin	suspensory ligament		

Cells called receptors detect stimuli (changes in the environment). Different parts of the body contain different receptors. Complete this table about receptors:

receptor cells	found in
light	1 _____ *e y e*
sound and balance	2 _____ *e a r*
temperature and pressure	3 _____ *s k i n*
chemicals	tongue and 4 _____ *n o s e*

The eye contains receptor cells. Label this drawing of a section through the eye:

5 _____
6 *Sclera*
7 *Ciliary   muscle*
8 *Suspensory   ligament*
9 *Cornea*
10 *Pupil*
11 *Lens*
Iris

12 _____
13 *Optic   nerve*
Retina

Jobs of parts of the eye

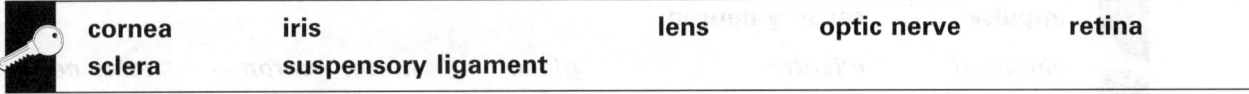

| cornea | iris | lens | optic nerve | retina |
| sclera | suspensory ligament | | | |

Complete the table about the jobs of parts of the eye.

part of eye	job
14 _____ _suspensory_ _____ _ligament_	attached to the lens, holds the lens in position
15 _____ _sclera_	the tough outer coat
16 _____ _iris_	controls the size of the pupil
17 _____ _retina_	contains receptor cells sensitive to light
18 _____ _cornea_	transparent to allow light to enter the eye
19 _____ _lens_	along with the cornea, produces an image on the retina
20 _____ _optic nerve_	contains sensory neurones that transmit impulses to the brain

Drugs

| addiction | brain | cancer | depressant | emphysema |
| liver | withdrawal symptoms | | | |

Solvents, tobacco smoke and alcohol can all affect our behaviour. The craving for drugs or solvents is

known as 21 _____ _addiction_ . People addicted to drugs suffer

22 _____ _withdrawal_ _____ _symptoms_ without them.

Sniffing solvents is most likely to cause damage to the lungs, the liver and the 23 _____ _brain_ .

Tobacco smoke contains substances which can cause lung 24 _____ _cancer_ and diseases of the

lungs such as 25 _____ _emphysema_ .

Alcohol affects the nervous system by slowing down our reactions and it is therefore known as

a 26 _____ _depressant_ .

In the long term alcohol abuse may cause damage to the brain and the 27 _____ _liver_ .

Transmission of information

> 🔑 impulse   sensory neuron
>
> *chemical*   *effector*   *gland*   *motor neuron*   *relay neuron*
>
> *response*   *synapse*

When stimulated, receptors produce nerve 28 _____ . These pass along nerve cells
impulses

called 29 _____ _____ towards the central nervous system where they form a
sensory   *neurons*

junction with a H30 _____ _____ .
relay   *neurons*

This junction is called a H31 _____ . The impulse is carried across the junction by
synapse

H32 _____ .
chemicals

Nerve cells that carry impulses away from the brain or spinal cord are called H33 _____
motor

_____ .
neurons

These neurons end in H34 _____ which are either muscles or
effectors

H35 _____ ; these bring about a H36 _____ .
glands   *response*

Control

> 🔑 *co-ordinator*   *effector*   *motor neuron*   *receptor*   *reflex action*
>
> *response*   *sensory neuron*   *stimulus*

Automatic control of an activity is called a H37 _____ _____ .
reflex   *action*

Narrowing the pupil in bright light is an automatic activity. Complete the table for this activity:

bright light	H38 _____ *stimulus*
cells in retina	H39 _____ *receptor*
brain	H40 _____ *co-ordinator*
muscle cells in iris	H41 _____ *effector*
pupils narrow	H42 _____ *response*

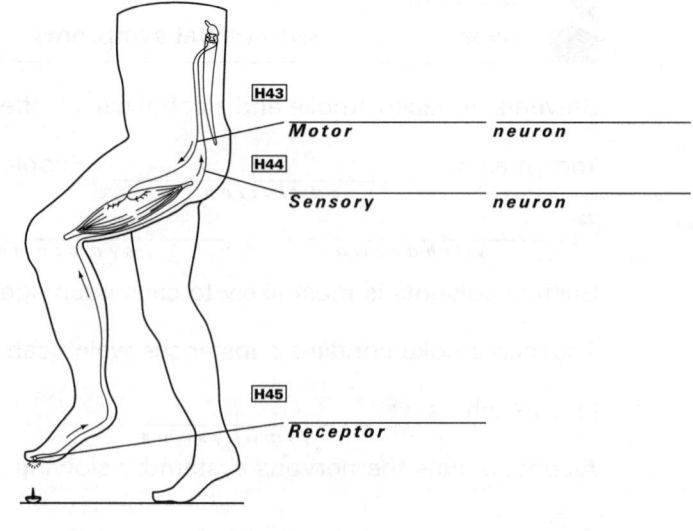

H43 _____
Motor   *neuron*

H44 _____
Sensory   *neuron*

H45 _____
Receptor

The drawing shows the structures in a

pain-withdrawal reflex. When the person

steps on the pin the muscle automatically

contracts. Label the structures in this

activity.

Focusing

| ciliary muscle | cornea | lens | retina | suspensory ligament |

Most refraction occurs at the junction between air and the [H46] _____ .
cornea

To focus on near objects the [H47] _____ _____ contract, lowering the tension
ciliary   muscles

in the [H48] _____ _____ .
suspensory   ligaments

The [H49] _____ becomes more convex and focuses the image on the
lens

[H50] _____ .
retina

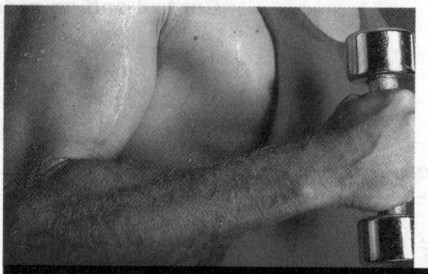

Homeostasis

Do all of this topic for Single Science and Double Science.
H = for Higher Tier only

📝 Revision notes

Waste materials

🔑	bladder	carbon dioxide	liver	lung	skin	sweat
	urea	urine				

The waste liquid produced by the skin is called [1] _____ .
sweat

The waste liquid produced by the kidneys is called [2] _____ .
urine

Complete this table showing where waste materials leave the blood:

	waste material
leaves the blood only in the lungs	[3] _____ _____ *carbon* *dioxide*
leaves the blood only in the kidneys	[4] _____ *urea*
leaves the blood in the skin, kidneys and [5] _____ *lungs*	excess water
leaves the blood in the kidneys and [6] _____ only *skin*	excess ions

The waste materials produced by aerobic respiration are water and [7] _____
carbon

_____ . The waste material produced by the breakdown of amino acids is
dioxide

[8] _____ . This breakdown takes place mainly in the [9] _____ .
urea   *liver*

Urine is produced in the kidneys and then stored in the [10] _____ until it leaves the body.
bladder

Excess heat is transferred from the body mainly via the [11] _____ .
skin

Science: On Course for GCSE NEAB Edition Teacher's Book © Stanley Thornes (Publishers) Limited 1998

Chemical co-ordination

diabetes	gland	glucagon	hormone	pancreas	plasma
glucagon	*glycogen*				

Chemical messengers are called [12] _____ . They are made in [13] _____ and
             hormones   *glands*

transported to their target organs by [14] _____ . Blood sugar levels are controlled by
        plasma

hormones produced by the [15] _____ . These hormones are called insulin and
      pancreas

[16] _____ . If insufficient insulin is made the disease [17] _____ results, one
   glucagon   *diabetes*

effect of which is high levels of glucose in the blood. The pancreas monitors blood sugar level. If blood

sugar level is too low it secretes a hormone called [H18] _____ which causes the liver to
                      glucagon

convert stored [H19] _____ into glucose. If blood sugar level is too high, the pancreas
       glycogen

produces insulin which causes the liver to convert excess glucose into glycogen.

How the kidneys work

ADH	filtration	sugar	urea	water

The first stage in urine production is [H20] _____ of blood under high pressure. As
              filtration

the filtrate flows through the kidney tubules all of the [H21] _____ is re-absorbed into the
              sugar

blood. The ions and [H22] _____ needed by the body are also re-absorbed. The remaining
      water

filtrate, called urine, consists mainly of excess water, excess ions and [H23] _____ .
                  urea

If the water content of the blood is too low the hormone [H24] _____ is secreted.
              ADH

This [H25] *increases*/~~decreases~~ the rate of water re-absorption in the kidneys.

Temperature regulation

constrict	dilate	evaporate	respiration	thermoregulatory centre

Body temperature is monitored and controlled by the [H26] _____
              thermoregulatory

_____ in the brain. If the body is too warm, blood vessels supplying the skin capillaries
centre

[H27] _____ to increase the blood flow through the capillaries. If the body is too cold these
    dilate

blood vessels [H28] _____ to reduce the flow of blood through the skin capillaries.
        constrict

Increased sweating cools the body as the sweat [H29] _____ . Shivering is the
              evaporates

contraction of muscles resulting in an increase in the rate of [H30] _____ in the
              respiration

muscle, releasing more heat.

Photosynthesis and growth

Do all of this topic for Double Science only.
H = for Higher Tier only

 ## Revision notes

Jobs of the parts of a plant

anchorage	photosynthesis	support	water

Add labels to this diagram to show the job of each part of the plant.

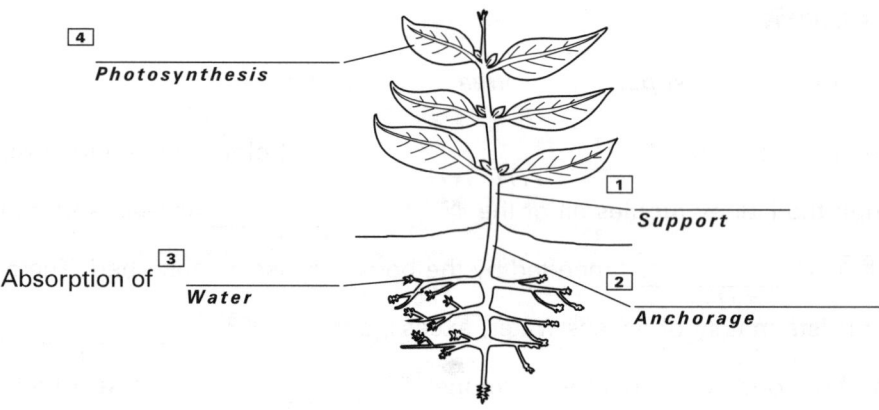

4 **Photosynthesis**

1 **Support**

3 Absorption of **Water**

2 **Anchorage**

The materials of photosynthesis

carbon dioxide	chlorophyll	chloroplast	glucose	light
oxygen	respiration	water		

Photosynthesis is the source of almost all the food we eat.

In order to photosynthesise, plants absorb 5 _____ _____ from the air and
carbon **dioxide**

6 _____ from the soil.
water

7 _____ energy is absorbed by a green pigment called 8 _____
light **chlorophyll**

which is found in the 9 _____ of leaf cells. The products of photosynthesis are a
chloroplasts

sugar called 10 _____ and the gas 11 _____ . This gas can be used in
glucose **oxygen**

12 _____ by both plants and animals.
respiration

Science: On Course for GCSE NEAB Edition Teacher's Book © Stanley Thornes (Publishers) Limited 1998

The cells which photosynthesise

🔑	cell membrane	cell sap	cell wall	chloroplast
	cytoplasm	nucleus	vacuole	

Label the cell from a green leaf.

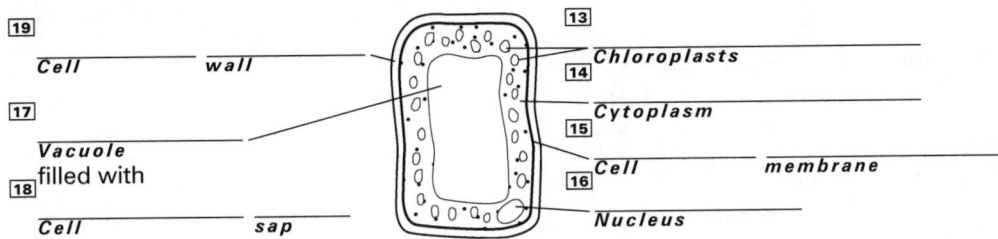

[19]
_Cell _____ wall_

[17]
Vacuole
[18] filled with

_Cell _____ sap_

[13]
Chloroplasts
[14]
Cytoplasm
[15]
_Cell _____ membrane_
[16]
Nucleus

The sugars produced by photosynthesis

🔑	carbon dioxide	chlorophyll	energy	growth	oxygen	starch
	cellulose	_nitrate_	_protein_			

Most of the sugars are used up in respiration to release [20] _____ .
energy

Some are converted into the many different molecules needed for the [21] _____ of young
growth

cells, including [H22] _____ which is needed to produce cell walls.
cellulose

Some are transported to other parts of the plant and stored as a carbohydrate called

[23] _____ . Others are combined with [H24] _____ to produce the
starch _nitrates_

[H25] _____ needed for growth and repair. The word equation for photosynthesis is:
protein

light [27] _____
energy

[26] _____ _____ + water ⟶ glucose + [29] _____
carbon _dioxide_ _oxygen_

[28] _____
chlorophyll

Factors that limit the rate of photosynthesis

🔑	carbon dioxide	light	temperature

Different factors limit the rate of photosynthesis at different times of the day and year.

At night on a summer's day the rate of photosynthesis is limited by low [30] _____ intensity.
light

In a closed greenhouse at noon on a summer's day the rate of photosynthesis is limited by low

[31] _____ _____ concentration. At noon on a sunny winter's day the rate of
carbon _dioxide_

photosynthesis is most limited by the low [32] _____ .
temperature

Science: On Course for GCSE NEAB Edition Teacher's Book © Stanley Thornes (Publishers) Limited 1998

Plant growth

🔑	cutting	fruit	gravity	hormone	light	root	water	weed

A bean seedling was placed on its side. Its appearance 48 hours later is shown below.

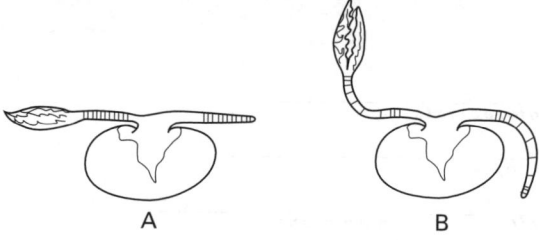

A   B

48 hours later

The shoot of the plant has grown against the force of [33] _____ and the root has grown in

gravity

the direction of this force. Most of the growth has occurred at the [34] ~~base~~/*tip* of the root. Growth in

plants is controlled and co-ordinated by [35] _____ . Plant stems grow towards the stimulus

hormones

of [36] _____ . Besides the effect of gravity, plant roots grow towards [37] _____ .

light   *water*

These responses are caused by [38] ~~equal~~/*unequal* distribution of [39] _____ in the roots or

hormones

stems.

Plant hormones can be used to kill [40] _____ in lawns or amongst crops.

weeds

A plant shoot that has been removed from the rest of the plant is known as a [41] _____ . If

cutting

this is dipped in hormone powder it will form [42] _____ . If hormones are applied to

roots

flowers they will control the development of [43] _____ .

fruits

Mineral requirements of plants

🔑	enzyme	limiting factor	nitrate	photosynthesis	protein

Complete this table of the uses of mineral ions in plants and the effects of their deficiency:

mineral ion	function	effect of deficiency
[H44] _____ *nitrate*	synthesis of [H45] _____ *protein*	stunted growth
phosphate	helps reactions involved in respiration and [H46] _____ to work *photosynthesis*	[H47] *purple*/~~yellow~~ leaves
potassium	helps [H48] _____ to work *enzymes*	[H49] ~~purple~~/*yellow* leaves with dead spots

In Third World countries mineral ions are often the principal [H50] _____ _____

limiting   *factor*

in reducing crop yields.

Water relations

Do all of this topic for Double Science only.
H = for Higher Tier only

 ## Revision notes

Path taken by water through a plant

| evaporate | root hair | stoma (plural stomata) | transpiration | xylem |

Label this drawing showing the path taken by water through a plant:

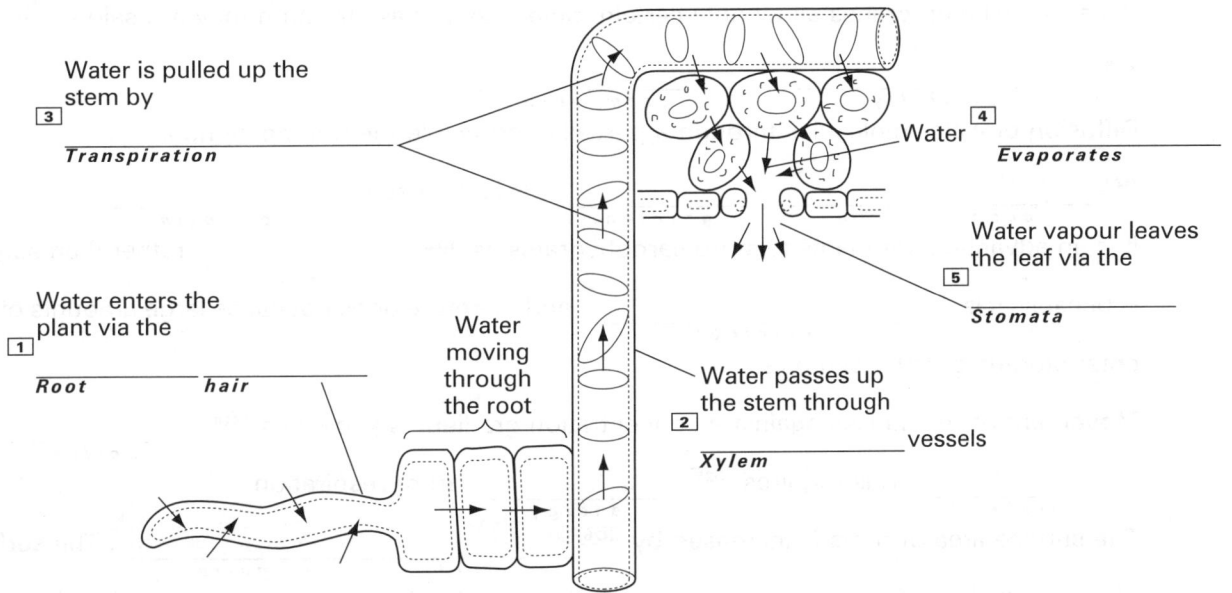

Water is pulled up the stem by
3
Transpiration

Water enters the plant via the
1
Root   hair

Water moving through the root

Water passes up the stem through
2
Xylem   vessels

Water   4
Evaporates

Water vapour leaves the leaf via the
5
Stomata

Loss of water vapour from leaves

| carbon dioxide | guard cells | transpiration | wax | wilting |

The loss of water through leaves is called ⑥ _____ . The rate of this loss is fastest
                                        transpiration

when external conditions are ⑦ *hot*/~~cold~~, ⑧ *windy*/~~still~~ and when humidity is ⑨ ~~high~~/*low*.

The surface of a leaf is covered by ⑩ _____ to reduce the rate of water vapour loss. In
                                    wax

plants living in dry conditions this layer is ⑪ *thicker*/~~thinner~~ than in other plants.

Science: On Course for GCSE NEAB Edition Teacher's Book © Stanley Thornes (Publishers) Limited 1998

Stomata are holes surrounded by [12] _____ _____ . Their principle function is
                                        guard   *cells*

to allow [13] _____ _____ to enter the leaf. In most plants they are more
                   carbon   *dioxide*

abundant on the [14] *upper*/*lower* surface of leaves where there is [15] *more*/*less* air movement and

where it is [16] *warmer*/*cooler*. If plants lose water faster than they can absorb it the stomata may close

to prevent the plant [17] _____ .
                              wilting

Movement of substances across boundaries

phloem

active uptake   air space   energy   partially permeable
concentration gradient insoluble   osmosis   root hair
starch

Carbohydrates are moved around the plant through a tissue called [18] _____ .
                                                                          phloem

Because cell membranes allow only certain molecules to pass through they are said to be

[H19] _____ _____ .
          partially   *permeable*

Diffusion of water molecules through a partially permeable membrane along a

[H20] _____ _____ is known as [H21] _____ .
         concentration   *gradient*   *osmosis*

It is an advantage for plants to store carbohydrates as [H22] _____ rather than sugars. This
                                                                   starch

is because it is [H23] _____ and therefore does not cause large amounts of water to
                          insoluble

enter storage cells by osmosis.

Movement of substances against a concentration gradient is known as [H24] _____
                                                                              active

_____ . This requires [H25] _____ from respiration.
   uptake   *energy*

The surface area of roots is increased by [H26] _____ _____ . The surface area
                                                     root   *hairs*

of leaves available for gaseous exchange is increased by their flattened shape and by internal

[H27] _____ _____ .
          air   *spaces*

Support

cell wall   osmosis   turgor

As water moves into a plant cell by [H28] _____ it increases the pressure on the
                                              osmosis

[H29] _____ _____ of the cell. This pressure is known as [H30] _____
          cell   *wall*   *turgor*

pressure and is the principle means of support for leaves and for young plants.

The environment

Do all of this topic for Single Science and Double Science.
H = for Higher Tier only

 ## Revision notes

Environmental factors

carbon dioxide	light	nutrient
oxygen	temperature	water

Label this diagram which shows the environmental factors that affect a tree.

[3] _____ energy from the Sun
Light

[4] _____ and
Oxygen
[5]
_____
Carbon   Dioxide
from the air

[1] W_____ and n _____ from the soil
Water   [2]   *Nutrients*

In addition to these factors trees also need a suitable [6] _____ so that the
temperature

chemical reactions inside their cells will occur at a reasonable rate.

Competition

breeding	nutrient	predator	prey

Plants compete with each other for light, water and nutrients. Animals compete with each other for

[7] _____ and space for [8] _____ .
nutrients   *breeding*

Animals that eat other animals are called [9] _____ ; the animals that are eaten are
predators

called [10] _____ . If the number of prey rises, the number of predators will usually
prey

[11] rise/~~fall~~; if the number of predators rises the number of prey will usually [12] ~~rise~~/fall.

Effects of increases in the size of human populations

| combustion | fossil fuel | non-renewable | pollution |

Increases in the size of human populations have led to increased use of [13] _____
n o n - r e n e w a b l e
energy resources such as [14] _____ _____ . The burning of fuels is known as
fossil *fuels*
[15] _____ ; this leads to [16] _____ of the air.
c o m b u s t i o n *pollution*

Air pollution

| acidic | carbon dioxide | leaf | sulphur dioxide |

Power stations affect the environment.

The most abundant gas in the smoke from power stations is [17] _____ _____ .
c a r b o n *d i o x i d e*
There are also significant amounts of the gas [18] _____ _____ which dissolves
s u l p h u r *d i o x i d e*
in rain making it [19] _____ .
a c i d i c
When this rain falls it causes the trees to lose some of their [20] _____ .
l e a v e s
When this rainwater reaches the lake it causes the lake to become [21] _____ , killing many
a c i d i c
of the organisms that live there.

Eutrophication

| competition | fertiliser | microbe | oxygen | pesticide | respiration |

Farmers use [H22] _____ to replace nutrients in the soil and
fertilisers
[H23] _____ to kill organisms which damage crops. If fertilisers or sewage reach fresh
pesticides
water, water plants grow rapidly and as a result many die through [H24] _____ for
competition
light. This results in an increase in the populations of [H25] _____ , which feed on dead
microbes
organisms, and their [H26] _____ reduces the [H27] _____ concentration
respiration *oxygen*
of the water, resulting in the death of many of the animals.

The greenhouse effect

| carbon dioxide | methane | radiation |

Increases in the number of cattle and in the number of rice fields have resulted in an increase in

the [H28] _____ content of the atmosphere.
methane
Deforestation has reduced the rate at which [H29] _____ _____ is removed
c a r b o n *d i o x i d e*
from the atmosphere and locked up in wood.

These two gases reduce the amount of energy lost from the Earth by H30 _____ ,
radiation
causing the mean temperature of the Earth to rise. This is known as the greenhouse effect.

Energy flow and nutrient cycles

Do all of this topic for Double Science only.
H = for Higher Tier only

Revision notes

Food chains

 consumer   energy   photosynthesis   producer   radiation

This diagram shows food chains for some of the organisms in an aquarium.

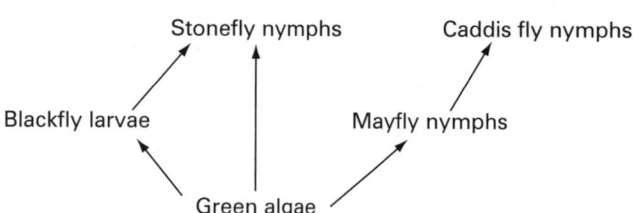

In the food chains, algae are ☐1 _____ . Only the algae can carry out

producers

☐2 _____ . All the other organisms eat food and are therefore known as

photosynthesis

☐3 _____ .

consumers

A food chain shows how ☐4 _____ is transferred between the organisms in a community.

energy

The source of all the energy in a food chain is ☐5 _____ from the sun.

radiation

Ecological pyramids

The trophic levels of a community can be arranged in a pyramid shape, with producers at the base.

Pyramids of numbers illustrate the number of organisms at each stage in a food chain. Pyramids of

biomass illustrate the total mass of biological materials at each stage in a food chain.

These diagrams show four ecological pyramids: A, B, C and D.

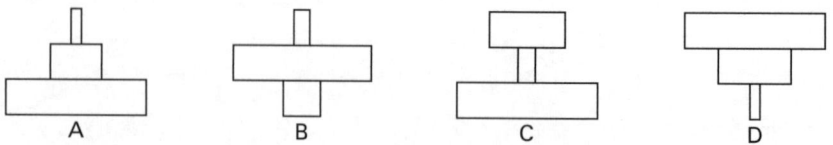

For the food chain: grass → rabbits → fox, the pyramid of numbers is ☐6 *A/B/C/D* and the pyramid of

biomass is ☐7 *A/B/C/D*.

Science: On Course for GCSE NEAB Edition Teacher's Book © Stanley Thornes (Publishers) Limited 1998

For the food chain: oak tree → insects → bird, the pyramid of numbers is [8] *A/B/C̶/D̶* and the pyramid of biomass is [9] *A/B̶/C̶/D*.

The pyramid of numbers for the food chain: grass → cow → fleas is [10] *A/B/C/D̶*.

Decay

🗝	compost heap	detritus feeder	digestion	microbe	oxygen
	sewage works				

When living organisms die their bodies may be eaten by animals called [11] _____ *detritus*
_____ or they may decay. Decay is the [12] _____ of dead material
feeders   *digestion*
by [13] _____ .
        microbes

Decay occurs fastest in [14] *moist/dry̶* and [15] *warm/cold̶* conditions. Most organisms that cause decay

also need [16] _____ for their respiration. Microbes are used to break down human faeces
              oxygen

in [17] _____ _____ and to break down waste plant materials in
        sewage   *works*

[18] _____ _____ .
     compost   *heaps*

The carbon cycle

🗝	carbohydrate	microbe	photosynthesis	respiration

Green plants remove carbon dioxide from the atmosphere for [19] _____ and
                                                               photosynthesis

convert it first into [20] _____ . Some of the carbon dioxide is returned to the
                           carbohydrate

atmosphere by the [21] _____ of the green plants.
                       respiration

When green plants are eaten by an animal some of the carbohydrate becomes part of the animal.

Some of this carbohydrate is used by animals in [22] _____ to release energy;
                                                      respiration

carbon dioxide released by this process returns to the atmosphere.

When animals and plants die some animals and [23] _____ feed on their bodies. Carbon
                                                   microbes

dioxide is returned to the atmosphere during the [24] _____ of these organisms.
                                                       respiration

Energy loss in food chains

🗝	*faeces*	*heat*	*movement*

At each stage in a food chain less energy and materials are contained in the biomass of the organisms.

Some materials and energy are lost in the organism's [H25] _____ . Animals need energy for
                                                           faeces

[H26] _____ and much of this energy is lost to the environment as [H27] _____ .
      movement   *heat*

Eventually all the energy captured by green plants is returned to the environment.

Science: On Course for GCSE NEAB Edition Teacher's Book © Stanley Thornes (Publishers) Limited 1998

The nitrogen cycle

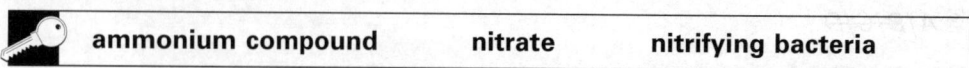

| 🔑 | ammonium compound | nitrate | nitrifying bacteria |

Label this diagram of the nitrogen cycle:

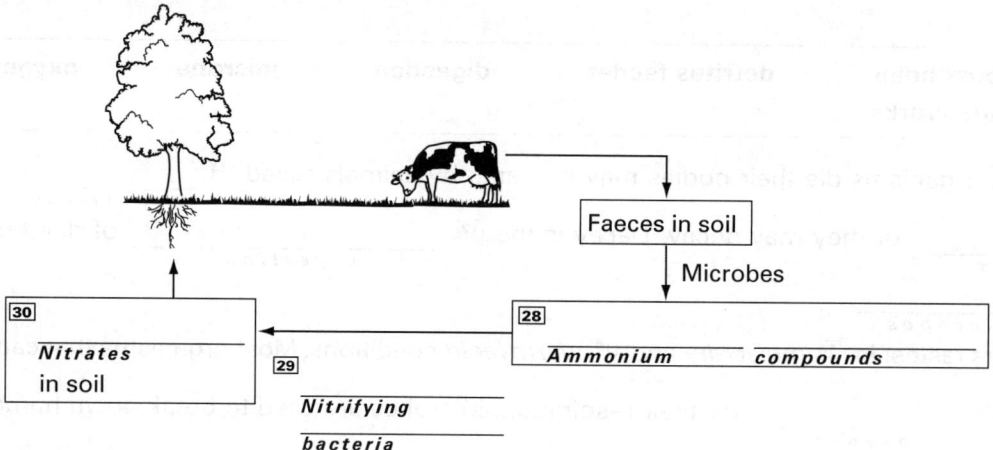

Faeces in soil

Microbes

| 30 |
| *Nitrates* |
| in soil |

| 29 |
| *Nitrifying* |
| *bacteria* |

| 28 |
| *Ammonium*   *compounds* |

Science: On Course for GCSE NEAB Edition Teacher's Book © Stanley Thornes (Publishers) Limited 1998

Variation and selection

Do all of this topic for Single Science and Double Science.
H = for Higher Tier only

 ## Revision notes

Variation

allele	asexual	cancer	chromosome	clone
environmental	fertilisation	gene	genetic	mutation
radiation	sexual			

The diagram shows some strawberries. They all belong to the same variety and they were all picked on the same day.

The strawberries are similar because they all contain similar [1] _____ . The
genes
differences in size are due to

[2] *e* _____ factors and
environmental
[3] *g* _____ factors.
genetic
New strawberry plants are produced when the parent plant forms runners. Runners are produced by cell division at the base of the parent cell. This type of reproduction is known as [4] _____ reproduction. Because of this the genetic information in the cells of the parent
asexual
plant and the young plant is [5] *identical*/~~different~~.

Strawberry fruits are produced when sex cells from two strawberry flowers fuse. The process by which two sex cells fuse is called [6] _____ and this method of reproduction is
fertilisation
said to be [7] _____ . The fruits produced by this method of reproduction contain
sexual
[8] ~~identical~~/different genetic information. Genetically identical organisms are called

[9] _____ .
clones

Science: On Course for GCSE NEAB Edition Teacher's Book © Stanley Thornes (Publishers) Limited 1998

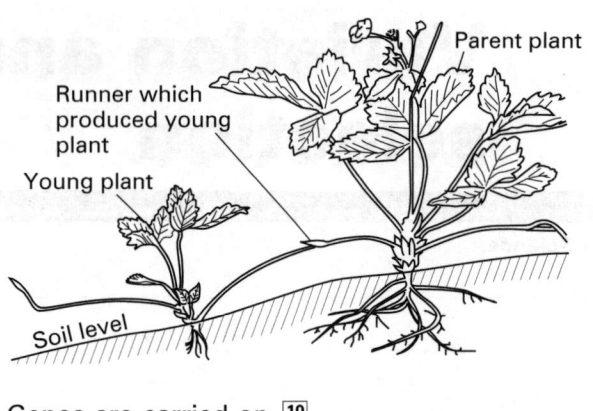

Parent plant

Runner which produced young plant

Young plant

Soil level

Genes are carried on [10] _____ chromosomes . Many genes have different forms called

[11] _____ alleles . New forms of genes result from changes, called [12] _____ mutations ,

to existing genes. The frequency of these changes is increased by exposure to ionising

[13] _____ radiation and to certain chemicals. Some mutations in body cells cause

uncontrolled cell division which may result in [14] _____ cancer .

Controlling inheritance

| artificial selection | allele | characteristic | cutting |

New plants can be produced quickly and cheaply by taking [15] _____ cuttings from the stems of

older plants. These plants are genetically [16] *identical/~~different~~* to the parent plant. We can use

[17] _____ artificial _____ selection to produce new varieties of plants and

animals. We do this by choosing breeding individuals that have [18] _____ characteristics that are

useful to us. One disadvantage of selective breeding is that it reduces the number of

[H19] _____ alleles in a population. Widespread use of cloning has the same effect.

Cell division

| *allele* | *meiosis* | *mitosis* | *parent* |

The cells of offspring produced by asexual reproduction are produced by [H20] _____ mitosis from

the parental cells. Sex cells are produced by [H21] _____ meiosis from the parental cells.

During meiosis the number of chromosomes in a cell [H22] *halves/~~doubles/stays the same~~*.

Sexual reproduction gives rise to variation because:

● gametes are produced by [H23] _____ meiosis ;

● when gametes fuse, one of each pair of genes comes from each [H24] _____ parent ;

● the genes in each pair may be different [H25] _____ alleles .

Science: On Course for GCSE NEAB Edition Teacher's Book © Stanley Thornes (Publishers) Limited 1998

Cloning and genetic engineering

	bacteria	DNA	embryo	protein	tissue culture

Modern cloning techniques include:

● using small groups of cells from part of a plant: this is known as [H26] _____
tissue

_____;
culture

● splitting apart cells from a developing young mammal and allowing these to develop in the womb

of an adult female mammal: this is known as [H27] _____ transplant.
embryo

The molecule that carries genetic information is called [H28] _____ . It carries this
DNA

information as a code for the sequence of amino acids in [H29] _____ . In genetic
proteins

engineering a sequence of this molecule is cut out from a host cell and transferred into

[H30] _____ which then divide rapidly and make large quantities of the substance coded for.
bacteria

Science: On Course for GCSE NEAB Edition Teacher's Book © Stanley Thornes (Publishers) Limited 1998

Inheritance and evolution

Do all of this topic for Single Science and Double Science.
H = for Higher Tier only

Revision notes

Genetic information

🔑 allele   chromosome   nucleus

Genetic information is located inside the ①_____ of a cell. Genes are carried on
nucleus

②_____ . Many genes have two forms called ③_____ . In body cells
chromosomes   *alleles*

the chromosomes are ④ *single*/in pairs whereas in gametes they are ⑤ single/*in pairs*.

Sex determination

In human body cells one pair of chromosomes carries the genes which determine sex. These are

called sex chromosomes and there are two types: X chromosomes and Y chromosomes.

Complete this diagram to show how sex is determined in humans. Use 'X' for an X chromosome and

'Y' for a Y chromosome.

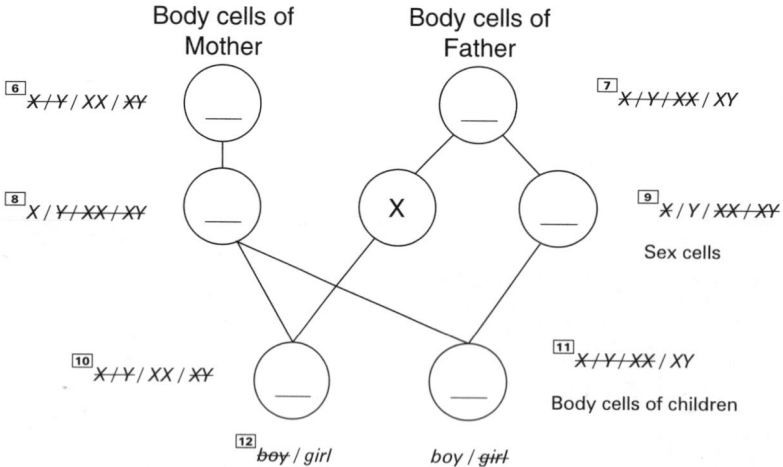

Genetic terms

🔑 *carrier*   *dominant*   *heterozygous*   *homozygous*   *recessive*

Science: On Course for GCSE NEAB Edition Teacher's Book © Stanley Thornes (Publishers) Limited 1998

An allele which controls the development of a characteristic when it is present on only one of the chromosomes is a [H13] _____ allele. An allele which controls the development of a
dominant
characteristic only when it is present on both of the chromosomes is a [H14] _____ allele. If
recessive
both chromosomes in a pair contain the same allele of a gene the individual is

[H15] _____ for that gene. If the chromosomes in a pair contain different alleles of a
homozygous
gene, the individual is [H16] _____ for that gene. Heterozygous individuals are said
heterozygous
to be [H17] _____ of diseases caused by recessive alleles.
carriers

Inheritance of a disorder caused by a dominant allele

Huntington's chorea is caused by a dominant allele H. Complete the checkerboard diagram to show the probability of a child inheriting the disorder from a heterozygous father (Hh) and a homozygous recessive mother (hh).

alleles in egg	alleles in sperm	
	H	[H18] *H/h*
h	[H20] *HH/Hh/h̶h̶*	[H21] *HH/Hh/hh*
[H19] *H/h*	[H22] *HH/Hh/h̶h̶*	[H23] *H̶H̶/Hh/hh*

The probability of the child inheriting the disorder is [H24] *none/1 : 1/1̶ ̶:̶ ̶2̶/1̶ ̶:̶ ̶4̶.*

Inheritance of a disorder caused by a recessive allele

Cystic fibrosis is caused by a recessive allele c. It can be passed on by parents even though neither of them has the disorder.

Complete the checkerboard diagram to show the probability of a child inheriting the disorder from a heterozygous father (Cc) and a homozygous dominant mother (CC).

alleles in egg	alleles in sperm	
	C	[H25] *C̶/c*
C	[H27] *CC/C̶c̶/c̶c̶*	[H28] *C̶C̶/Cc/c̶c̶*
[H26] *C/c̶*	[H29] *CC/C̶c̶/c̶c̶*	[H30] *C̶C̶/Cc/c̶c̶*

The probability of the child inheriting the disorder is [H31] *none/2̶5̶%̶/5̶0̶%̶/7̶5̶%̶/1̶0̶0̶%̶.*

Science: On Course for GCSE NEAB Edition Teacher's Book © Stanley Thornes (Publishers) Limited 1998

Hormonal control of the menstrual cycle

contraceptive drug	fertility drug	hormone	pituitary	womb
FSH	*LH*	*oestrogen*	*pituitary*	

The monthly release of an egg from a woman's ovaries and changes in the thickness of the lining of

her [32] _____ (womb) are controlled by chemicals called [33] _____ (hormones). These hormones

are produced by a gland at the base of the brain called the [34] _____ (pituitary) gland and by

the ovaries themselves.

Fertility in women can be controlled by giving them either:

- [35] _____ _____ (fertility) (drug) to stimulate the release of eggs, or

- [36] _____ _____ (contraceptive) (drugs) to prevent the release of eggs.

- The hormone produced by the pituitary gland which causes eggs in the ovary to mature is

  called [H37] _____ (FSH).

- This hormone also stimulates the ovaries to produce hormones called [H38] _____ (oestrogens).

- The hormones produced by the ovaries inhibit the production of [H39] _____ (FSH) but

  stimulate the production of another hormone called [H40] _____ (LH) which is produced by

  the [H41] _____ (pituitary) gland.

- The hormone which stimulates the release of an egg is called [H42] _____ (LH).

Fertility drugs contain the hormone [H43] _____ (FSH) to stimulate the maturation of eggs in the

ovaries. Contraceptive drugs contain the hormone [H44] _____ (oestrogen) which inhibits the

production of the hormone [H45] _____ (FSH).

Evolution

decay	evolution	fossil
adaptation	*variation*	

The remains of organisms which lived many years ago are called [46] _____ (fossils). They are most

often formed from parts of organisms that do not [47] _____ (decay). Most of the evidence for

[48] _____ (evolution) comes from a study of the way in which organisms have changed over
long periods of time.

The theory of evolution by natural selection states that all members of a species show

[H49] _____ (adaptations) and that some of these become [H50] _____ (variations) that

enable the organism to survive changing conditions, whereas other organisms become extinct.

Science: On Course for GCSE NEAB Edition Teacher's Book © Stanley Thornes (Publishers) Limited 1998

Metals

Do part of this topic for Single Science and all of it for Double Science.
D = for Double Science only   H = for Higher Tier only

📝 Revision notes

Metals and non-metals

🔑 | boiling point | carbon | metal | soft |

An element which is shiny, conducts heat and electricity and has a high density is a

[1] _____ .
    metal

Non-metals have low melting points and [2] _____ _____ . They are dull and
                 boiling   *points*

are mostly [3] _____ and crumbly when solid. They are usually poor conductors of heat
          soft

and electricity. One non-metal which is a good conductor of electricity is [4] _____ .
                                          carbon

Metals and the reactivity series

🔑 | copper | displacement | hydrogen | hydroxide | iron |
| magnesium | oxide | oxygen | paraffin oil | reactivity |
| salt | zinc | | | |

Metals are arranged in order of [5] _____ with the most reactive metals at
                                   reactivity

the [6] ~~bottom~~/top of the list. Metals at the top of the list, such as potassium and sodium, are stored

under [7] _____ _____ to prevent them reacting with [8] _____ ,
          paraffin   *oil*   *oxygen*

water and carbon dioxide in the air.

When a metal reacts with oxygen in the air it will form an [9] _____ . When a metal reacts
                                                    oxide

with water (or steam) it will form a metal [10] _____ and [11] _____ gas.
                                    hydroxide   *hydrogen*

If a metal reacts with a dilute acid, such as hydrochloric acid, a metal [12] _____ is formed
                                                               salt

and [13] _____ gas. One metal which does not react with water or dilute hydrochloric acid
          hydrogen

is [14] _____ .
      copper

Reactions of metals can be predicted using the reactivity series. If a metal is added to a less reactive

metal which is in a compound, a reaction will take place. This type of reaction is called a

[15] _____ reaction.
    displacement

Science: On Course for GCSE NEAB Edition Teacher's Book © Stanley Thornes (Publishers) Limited 1998

If iron filings are added to blue copper(II) sulphate solution, a brown deposit of

[16] _____ and a pale green solution of [17] _____ (II) sulphate are formed.
     copper   *iron*

Two metals which will displace lead from a solution of lead(II) nitrate are [18] z_____ and
                                                                                zinc

[19] m_____ .
     magnesium

Extraction of metals

| electrolysis | reduction | rock | uncombined |

The photograph shows ore being dug out of the ground.

An ore is a [D20] _____ containing a metal or compounds of a metal.
              rock

Metals high in the reactivity series, such as sodium and aluminium, are extracted from their ores by

[D21] _____ . Metals in the middle of the reactivity series, such as zinc and iron, are
      electrolysis

extracted by [D22] _____ . Metals low in the reactivity series, such as gold, may be
                   reduction

found [D23] _____ in the Earth.
            uncombined

Extraction of iron

| acidic impurity | air | aluminium oxide | blast | carbon |
| carbon dioxide | coke | reduced | reducing agent | slag |

Iron is extracted from iron ore in a [D24] _____ furnace. Iron ore, [D25] _____ and
                                            blast   *coke*

limestone are loaded into the furnace. Blasts of hot [D26] _____ are blown into the furnace.
                                                            air

Coke is a form of the element called [D27] _____ . Coke burns to form [D28] _____
                                            carbon   *carbon*

_____ which is reduced by more coke to form carbon monoxide.
 dioxide

Iron oxide in the ore is [D29] _____ to iron by carbon monoxide. Carbon monoxide is the
                               reduced

[D30] _____ _____ . Calcium oxide, formed by the decomposition of calcium
      reducing   *agent*

carbonate (limestone), reacts with [D31] _____ _____ in the iron ore to
                                          acidic   *impurities*

form [D32] _____ .
           slag

Another way of producing iron from iron(III) oxide is to heat a mixture of aluminium powder and

iron(III) oxide. In this reaction aluminium is the [D33] _____ _____ . The products
                                                         reducing   *agent*

are iron and [D34] _____ _____ .
                   aluminium   *oxide*

Extraction of aluminium

| aluminium | burn | carbon | carbon dioxide |
| cryolite | electrolysis | oxygen |

Science: On Course for GCSE NEAB Edition Teacher's Book © Stanley Thornes (Publishers) Limited 1998

Aluminium is extracted from purified aluminium oxide by [D35] _____ .
electrolysis

Pure aluminium oxide is dissolved in molten [D36] _____ . The electrodes are made of
cryolite

[D37] _____ . An electric current is passed through the electrolyte. The product at the
carbon

negative electrode is [D38] _____ . The element produced at the positive electrode
aluminium

is [D39] _____ .
oxygen

The positive electrodes have to be replaced often because they [D40] _____ to produce
burn

[D41] _____ _____ .
carbon *dioxide*

Purification of copper

| electrolysis | electrolyte | ion | negative electrode | positive electrode |

oxidation *reduction*

The diagram shows the purification of impure copper by [42] _____ .
electrolysis

Positive electrode   Negative electrode

Copper(II) sulphate solution

Copper(II) sulphate solution in this process is called the [D43] _____ .
electrolyte

Copper is produced at the [D44] _____ _____ .
negative *electrode*

Copper at the [D45] _____ _____ dissolves in the copper(II) sulphate
positive *electrode*

solution to replace the copper deposited.

Copper(II) sulphate solution contains positive and negative [D46] _____ .
ions

At the negative electrode positively charged ions [DH47] *gain*/~~lose~~ electrons. This is called

[DH48] _____ .
reduction

At the positive electrode negatively charged ions [DH49] ~~gain~~/*lose* electrons. This is called

[DH50] _____ .
oxidation

Science: On Course for GCSE NEAB Edition Teacher's Book © Stanley Thornes (Publishers) Limited 1998

Acids, bases and salts

Do all of this topic for Single Science and Double Science.

📝 Revision notes

Acids and alkalis

🔑 | acid | alkali | hydrogen | neutral | salt | sour

The photograph shows substances which contain [1] _____ . They have a
a c i d s
[2] _____ taste. Acids contain the element [3] _____ which is replaced by a
s o u r   *h y d r o g e n*
metal when the acid is turned into a [4] _____ . Substances with pH values less than 7 are
s a l t
[5] _____ . Substances with pH values greater than 7 are [6] _____ .
a c i d s   *a l k a l i s*
A substance with a pH value of exactly 7 is [7] _____ .
n e u t r a l

pH

🔑 | meter | strong | universal | weak

The pH of a solution can be found by using [8] _____ indicator or a pH
u n i v e r s a l
[9] _____ . A solution with a pH value of 8 is a [10] _____ alkali and a solution
m e t e r   *w e a k*
with a pH value of 1 is a [11] _____ acid.
s t r o n g

Reactions of acids

🔑 | carbon dioxide | hydrogen | limewater | salt | sulphate

When magnesium ribbon is added to an acid, bubbles of colourless [12] _____ gas are
h y d r o g e n
seen. When sodium carbonate crystals are added to an acid, bubbles of colourless

[13] _____ _____ gas are seen. This gas turns [14] _____
c a r b o n   *d i o x i d e*   *l i m e w a t e r*
milky.

When dilute sulphuric acid is warmed with black copper(II) oxide, a blue solution of copper(II)

[15] _____ is formed.
s u l p h a t e
Mixing an alkali with an acid produces water and a [16] _____ .
s a l t

Salts

🔑	chloride	neutralisation	precipitation	water

A reaction between an acid and an alkali is called a [17] _____ reaction. The
neutralisation

products of this type of reaction are a salt plus [18] _____ .
water

When solutions of silver nitrate and sodium chloride are mixed [19] _____ takes
precipitation

place. The white solid is silver [20] _____ .
chloride

Equations

🔑	ammonia	carbon dioxide	nitric	sodium
	sulphate	sulphuric	water	

Complete the following word equations.

Sodium hydroxide + [21] _____ acid → [22] _____ nitrate + water
nitric   *sodium*

Zinc oxide + [23] _____ acid → zinc [24] _____ + [25] _____
sulphuric   *sulphate*   *water*

[26] _____ + sulphuric acid → ammonium sulphate
ammonia

Calcium carbonate + hydrochloric acid → calcium chloride + [27] _____ _____
carbon   *dioxide*

+ water

Zinc + [28] _____ acid → zinc sulphate + hydrogen
sulphuric

Lead carbonate + [29] _____ acid → lead nitrate + [30] _____ + carbon dioxide
nitric   *water*

Science: On Course for GCSE NEAB Edition Teacher's Book © Stanley Thornes (Publishers) Limited 1998

Rocks in the Earth

Do all of this topic for Double Science only.

Revision notes

Minerals and rocks

calcium carbonate	crystal	diamond	extrusive	igneous
intrusion	intrusive	magma	marble	metamorphic
mineral	sedimentary	shale	slate	

Although rocks are almost pure substances, most rocks are made up of a mixture of chemicals called [1] _____ *minerals* . The hardest mineral is [2] _____ *diamond* .

Rocks can be divided into three types. Rocks which are formed when liquid [3] _____ *magma* cools and crystallises are called [4] _____ *igneous* rocks. Two examples are granite and basalt. Granite is made up of large [5] _____ *crystals* formed when the magma cools [6] ~~quickly~~/slowly inside the Earth. A rock which crystallises inside the Earth rather than outside is called an

[7] _____ *intrusive* rock. Basalt is made up of small [8] _____ *crystals* formed when the magma cools [9] quickly/~~slowly~~ on the surface of the Earth. Rocks such as basalt, formed on the surface of the Earth, are called [10] _____ *extrusive* rocks.

Rocks which are formed when sediments are deposited and compressed are called

[11] _____ *sedimentary* rocks. The photograph shows a chalk cliff. Different layers can be clearly seen. Rocks in the lower layers are [12] ~~younger~~/older than the rocks above them.

A sedimentary rock formed from very fine particles is called [13] _____ *shale* or mudstone. It is a [14] ~~hard~~/soft rock.

When sedimentary or igneous rocks are subjected to [15] high/~~low~~ pressures and [16] high/~~low~~ temperatures, they may change and form [17] _____ *metamorphic* rocks. In Italy, for example, limestone has been turned into the metamorphic rock [18] _____ *marble* which is used to make statues or for facing buildings. Both limestone and marble are forms of the chemical compound

^[19] _____ _____. A metamorphic rock which is produced from
     c a l c i u m   *c a r b o n a t e*

mudstone is ^[20] _____ .
                s l a t e

The structure shown in this diagram is called an ^[21] _____ .
                                           i n t r u s i o n

Rocks at point **X** are ^[22] _____ rocks.
                             i g n e o u s

Rocks at point **Y** are ^[23] _____ rocks.
                             s e d i m e n t a r y

Rocks at point **Z** are ^[24] _____ rocks.
                             m e t a m o r p h i c

Rock cycle

🗝	burial	cementation	crystallisation	deposited	deposition
	erosion	extrusive	igneous	intrusive	magma
	melting	metamorphic	recrystallisation	sedimentary	transported
	weathering				

This diagram shows the rock cycle. Label this diagram using key words.

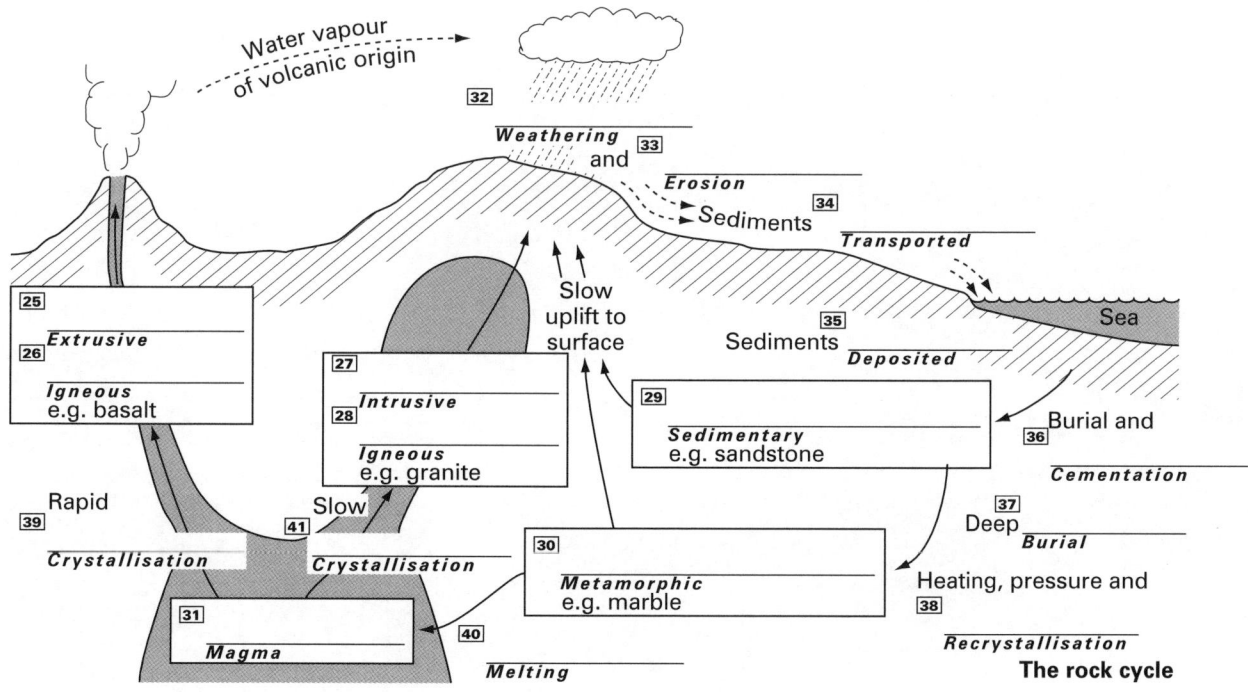

The rock cycle

Science: On Course for GCSE NEAB Edition Teacher's Book © Stanley Thornes (Publishers) Limited 1998

The steps involved in turning an igneous rock into a sedimentary rock would require the following processes (in the correct order):

42 _____ ,
      erosion

43 _____ ,
  transportation

44 _____ ,
     deposition

45 _____ ,
    burial

46 _____ .
   cementation

Fossils

| fossil | igneous | radioactivity | sedimentary |

Fossils are the remains of dead plants and animals which have been trapped in rocks over long periods of time. Fossils can be found in 47 _____ or metamorphic rocks but
          sedimentary

never in 48 _____ rocks.
   igneous

The presence of 49 _____ can be used to date rocks. Another way of dating rocks is to
     fossils

take measurements of 50 _____ .
     radioactivity

Science: On Course for GCSE NEAB Edition Teacher's Book © Stanley Thornes (Publishers) Limited 1998

Chemicals from oil

Do all of this topic for Single Science and Double Science.
H = for Higher Tier only

📝 Revision notes

Carbon compounds

🔑 | gas   organic   petroleum

Compounds of carbon, excluding simple compounds such as carbon dioxide and compounds such as

sodium carbonate, are called ☐1 _____ compounds. Crude oil
                                        organic

(or ☐2 _____) is a mixture of complex hydrocarbons found in the Earth. It was
                petroleum

formed over millions of years when high temperatures and pressures acted on the remains of tiny sea

creatures. It is often found with natural ☐3 _____ .
                                            gas

Fossil fuels

🔑 | carbon   fossil   gas   hydrocarbon
       non-porous rocks   oxygen   water

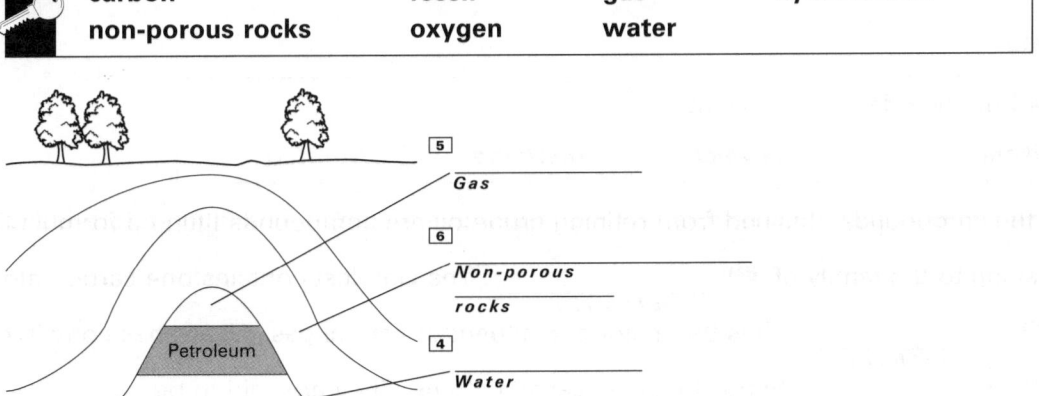

Label this diagram showing where oil and gas are found in the Earth.

Crude oil, natural gas and coal are all ☐7 _____ fuels. These fuels were formed from living
                                             fossil

materials which decayed in the absence of ☐8 _____ under high temperatures and
                                          oxygen

pressures. These fuels contain the element ☐9 _____ either as the element or combined.
                                          carbon

Compounds of carbon and hydrogen only are called ☐10 _____ .
                                                    hydrocarbons

Science: On Course for GCSE NEAB Edition Teacher's Book © Stanley Thornes (Publishers) Limited 1998

Oil refining

bitumen	boiling point	crude oil vapour	diesel
fractional distillation	lubricating oils	petrol	petroleum gas

Crude oil is refined by a process of [11] _____ _____ . This
fractional *distillation*

involves splitting up the mixture into different fractions, each fraction having a different range of

[12] _____ _____ . The diagram shows a column used for refining crude oil.
boiling *points*

[14]
Petroleum gases

[15]
Petrol

[16]
Diesel

[13]
Crude *oil* *vapour*

[17]
Lubricating oil

[18]
Bitumen

Label this diagram.

Fractions with a range of low boiling points condense near the [19] *top* / ~~*middle*~~ / ~~*bottom*~~ of the tower. As

you go down the tower the fractions have [20] *higher* / ~~*lower*~~ boiling point ranges.

Alkanes

carbon dioxide	viscosity		
alkane	*covalent*	*methane*	*saturated*

Most of the compounds obtained from refining crude oil are compounds fitting a formula C_nH_{2n+2}.

These belong to the family of [H21] _____ . The simplest contains one carbon atom and is
alkanes

called [H22] _____ . It is the major constituent of natural gas. All alkanes contain only
methane

single [H23] _____ bonds between carbon atoms. They are said to be
covalent

[H24] _____ compounds.
saturated

Alkanes are generally unreactive compounds. As the number of carbon atoms in the alkane increases,

the boiling point of the alkane [25] *increases* / ~~*decreases*~~, the resistance to pouring or

[26] _____ increases and the alkane is [27] ~~*easier*~~ / *more difficult* to burn.
viscosity

Hydrocarbons burn in air or oxygen. If they are burned in an excess of air, [28] _____
carbon

_____ and water are produced.
dioxide

Science: On Course for GCSE NEAB Edition Teacher's Book © Stanley Thornes (Publishers) Limited 1998

Cracking hydrocarbons

catalyst	cracking	
alkene	*ethene*	*unsaturated*

High boiling point fractions are of less economic value. They are often split up by a process

called [29] _____ into smaller molecules. This process involves passing the vapour of the
      cracking

high boiling point fraction over a heated [30] _____ . The smaller molecules produced often
                                              catalyst

contain hydrocarbons containing a double bond between carbon atoms. These are called

[H31] _____ compounds. They belong to the family of hydrocarbons called
        unsaturated

[H32] _____ .
        alkenes

The simplest alkene has a formula C_2H_4 and is called [H33] _____ .
                                                                ethene

Polymers

electricity	polyvinyl chloride	vinyl chloride	
monomer	*polymerisation*	*polymer*	*poly(propene)*

Small unsaturated molecules can be joined together by a process of addition

[H34] _____ to form long chains called [H35] _____ . The small molecules
        polymerisation   *polymers*

are called [H36] _____ . Joining together small molecules of propene produces a polymer
                  monomers

called [H37] _____ . PVC or [38] _____ _____ is a
              poly(propene)   *poly(vinyl*   *chloride)*

polymer made by joining together small molecules of [39] _____ _____ .
                                                           vinyl   *chloride*

PVC is used for covering electricity cables because it is a poor conductor of [40] _____ .
                                                                                    electricity

Science: On Course for GCSE NEAB Edition Teacher's Book © Stanley Thornes (Publishers) Limited 1998

The Earth and its atmosphere

Do all of this topic for Double Science only.
H = for Higher Tier only

✎ Revision notes

Structure of the Earth

convection	crust	force	inner core
liquid	mantle	nickel	outer core

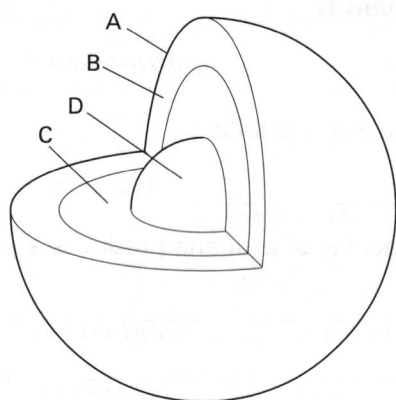

The diagram shows the Earth's structure. The layer labelled **A** is the very thin layer of surface rock called the [1] _____ . The layer labelled **B** lies beneath the crust and is called the
crust

[2] _____ . There is movement by [3] _____ in the mantle. The labels **C**
mantle   *convection*

and **D** represent the [4] _____ _____ and the [5] _____
outer   *core*   *inner*

_____ respectively. Both are made of the metals iron and [6] _____ . The inner
core   *nickel*

core is solid and the outer core is [7] _____ . The average density of the rocks in the crust
liquid

and mantle is less than the average density of the Earth, showing that the core is [8] *less dense*/*more*

dense than the rocks in the crust.

At the surface of the Earth younger sedimentary rocks lie on top of older rocks. These rock layers are

often tilted, folded, fractured and even turned upside down. This shows the Earth's crust is unstable

and has been subjected to very large [9] _____ .
forces

Science: On Course for GCSE NEAB Edition Teacher's Book © Stanley Thornes (Publishers) Limited 1998

Plate tectonics

🔑 | fossil | plate | tectonics |
boundary | continental | fault | magnetic field | mantle | oceanic
slide | subduction | volcano

The outer surface of the Earth is split into seven rigid sections or [10] _____ , each about
plates

100 km thick. Over time the plates have moved apart and the continents we know have been formed.

Evidence for this theory of plate [11] _____ has been obtained from the shapes of
tectonics

the continents, from rock structures and from [12] _____ .
fossils

Plates can move in three different ways. The photograph shows the effects of an earthquake. This

occurs when two plates [H13] _____ past each other. Stresses build up along cracks or
slide

[H14] _____ . These stresses are released when an earthquake occurs. There are a few
faults

earthquakes at the centres of plates, but most earthquakes occur at plate [H15] _____ .
boundaries

When two plates move apart new rocks are formed as rock from the [H16] _____ comes to
mantle

the surface. These new rocks, rich in iron, record the direction of the Earth's [H17] _____
magnetic

_____ and these patterns support the concept of the sea floor spreading.
field

When two plates move together the weaker [H18] _____ crust is forced underneath the
oceanic

[H19] _____ crust, returning rocks to the [H20] _____ . This is called
continental | *mantle*

[H21] _____ . The continental crust is forced upwards. Earthquakes are produced
subduction

and magma may rise through the continental crust to form [H22] _____ .
volcanoes

The Earth's atmosphere

🔑 | burned | global warming | greenhouse effect | mixture | oxygen |
photosynthesis | respire

Composition of the Earth's atmosphere by volume	
Nitrogen	78.0%
Oxygen	20.1%
Argon	0.9%
Carbon dioxide	0.03%
Neon	0.00015%
Other noble gases	0.0001%
Water vapour	variable

The air is a [23] _____ *mixture* of gases. The reactive gas in the air is [24] _____ *oxygen* . The composition of air remains approximately constant because some processes use up oxygen and produce carbon dioxide while other processes use up carbon dioxide and produce oxygen. Oxygen is used up and carbon dioxide is produced when plants and animals [25] _____ *respire* or when fuels containing carbon are [26] _____ *burned* . Green plants remove carbon dioxide from the air and replace it with oxygen in the process of [27] _____ *photosynthesis* .

Over the last century the levels of carbon dioxide in the atmosphere have [28] ~~decreased~~/increased/ ~~remained the same~~ because of the increasing amounts of carbon-rich fuels being burned and the destruction of forests. The build-up in levels of carbon dioxide is increasing the [29] _____ _____ *greenhouse* *effect* and may lead to [30] _____ _____ *global* *warming* .

Rates of chemical reactions

Do part of this topic for Single Science and all of it for Double Science.
D = for Double Science only

 ## Revision notes

Rates of reactions

activation energy	catalyst	chlorine	collision	compound
concentrated	cooling	cost	effective	explosion
pressure	reactant	start	surface	surface area
temperature				

The photograph shows a chemical reaction which is finished in a fraction of a second. It is a [1] *very fast*/*very slow* reaction. It is called an [2] _____ *explosion* . The erosion of stonework on a building is a [3] *very fast*/*very slow* reaction. As the time for a reaction increases, the rate of the reaction [4] *decreases*/*increases*/*stays the same*. Increasing the rate of reaction is important to industry as it reduces [5] _____ *costs* .

Rates of reaction can be explained using [6] _____ *collision* theory. A reaction takes place when a collision occurs with enough energy. This is called an [7] _____ *effective* collision and the amount of energy required is called the [8] _____ *activation* _____ *energy* .

Chemical reactions can be speeded up in a number of ways.

Increasing the temperature of a reaction will [9] *slow down*/*speed up* a chemical reaction. As a general rule, a temperature rise of 10 °C will often [10] *double*/*halve* the time for the reaction and so will [11] *halve*/*double* the rate of reaction. Heating up the reaction mixture [12] *speeds up*/*slows down* the particles. This will produce more effective [13] _____ *collisions* .

The chemical reactions which cause food to go bad can be slowed down by [14] _____ *cooling* the food in a refrigerator. Potatoes cook faster in hot fat than in boiling water because the fat is at a higher [15] _____ *temperature* than boiling water.

Science: On Course for GCSE NEAB Edition Teacher's Book © Stanley Thornes (Publishers) Limited 1998

Increasing the concentration of one or more of the reactants will [16] *increase*/decrease the time taken

for the reaction and so will [17] *increase*/~~decrease~~ the rate of reaction. In the case of a mixture of gases

reacting, increasing the [18] _____ *pressure* of the gases is the same as increasing the

concentrations. Increasing the concentration results in more [19] _____ *collisions* between

reacting particles and more of these will be effective leading to a faster reaction.

A solid in the form of a powder reacts [20] *faster than*/~~slower than~~/~~at the same rateas~~ the same mass

of the solid in the form of lumps. This is because the powder has a larger [21] _____ *surface*

_____ *area* to come in contact with the other reactant. Although coal does not readily catch

light, mixtures of coal dust and air can cause an [22] _____ *explosion* in a coal mine.

A mixture of hydrogen and chlorine reacts only very slowly but explodes in sunlight. Sunlight provides

energy to [23] _____ *start* the reaction by breaking some of the bonds between pairs of

[24] _____ *chlorine* atoms.

The rate of a chemical reaction can be changed by using a [25] _____ *catalyst* whose mass [26] *stays*

the same/~~increases~~/~~decreases~~ throughout the reaction. Using a catalyst produces [27] ~~more~~

~~product~~/~~less product~~/*the same mass of product* at the end of the reaction. A catalyst can work in

different ways. It can provide a [28] _____ *surface* where the reaction can take place or can form an

intermediate chemical [29] _____ *compound* which breaks down to give the products. Either way, it

lowers the [30] _____ *activation* _____ *energy* for the reaction so more collisions are

successful.

The diagram shows a rate of reaction experiment using calcium carbonate lumps and dilute

hydrochloric acid.

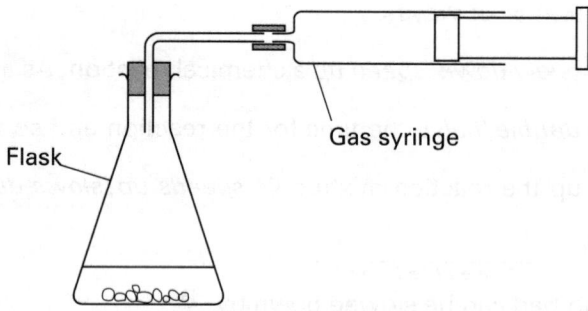

In this experiment the volume of gas collected is measured at intervals.

The graph shows the results obtained.

Science: On Course for GCSE NEAB Edition Teacher's Book © Stanley Thornes (Publishers) Limited 1998

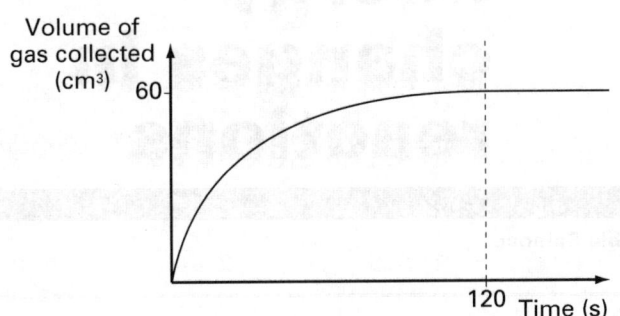

The graph is level when the reaction is [31] ~~fast/slow~~/finished and one of the

[32] _____ has been used up. In the early stages of the reaction the graph is
        reactants

steepest as the reaction is [33] fastest/~~slowest~~ at this stage because the acid is most

[34] _____ . If the experiment was repeated with the same mass of powdered
        concentrated

calcium carbonate, the volume of carbon dioxide collected would be [35] *the same*/~~more than~~

~~60 cm³/less than 60 cm³~~. The reaction is finished in [36] less than/~~more than/exactly~~ two minutes.

Enzymes

alcohol	amylase	carbon dioxide	denatured	enzyme
fermentation	irreversibly	lactic acid	lactose	limewater
saliva	yoghurt			

Biological catalysts are called [D37] _____ . Enzymes only operate within a certain range of
                                        enzymes

temperatures. At temperatures up to about 40 °C enzymes generally work [D38] faster/~~slower~~ with

increasing temperature. At high temperatures the protein molecules in the enzyme are

[D39] _____ and no reaction is possible. This is because the shape of the enzyme
        denatured

molecule is changed [D40] _____ . The molecules cannot fit into the site in the
                            irreversibly

enzyme molecule so no reaction is possible.

The enzyme used to split starch into maltose is called [D41] _____ . It is present in
                                                              amylase

[D42] _____ . Yeast cells turn sugar into [D43] _____ and [D44] _____
        saliva   *alcohol*   *carbon*

_____ gas. This process is [D45] _____ . The gas which turns
  dioxide   *fermentation*

[D46] _____ milky is carbon dioxide. Bread rises because bubbles of
        limewater

[D47] _____ _____ are trapped in the dough.
        carbon   *dioxide*

Bacteria use enzymes to turn milk into [D48] _____ by converting
                                              yoghurt

the [D49] _____ in the milk into [D50] _____ _____ .
            lactose   *lactic*   *acid*

Science: On Course for GCSE NEAB Edition Teacher's Book © Stanley Thornes (Publishers) Limited 1998

Energy changes in reactions

Do all of this topic for Single Science and Double Science.
H = for Higher Tier only

Revision notes

Burning

| carbon | energy | fuel | hydrogen | oxidation | oxygen |

The photograph shows a large forest fire. Burning is an [1] o_____ reaction which
oxidation

releases a large amount of [2] _____ to the surroundings. The trees are the
energy

[3] _____ . Burning involves the reaction with [4] _____ . Wood burns to produce
fuel   *oxygen*

carbon dioxide and water. This means that wood **must** contain the two elements

[5] c_____ and [6] _____ .
carbon   *hydrogen*

Exothermic and endothermic reactions

| endothermic | exothermic |

When a chemical change takes place energy changes may be observed. Reactions which give out

energy to the surroundings are called [7] _____ reactions and reactions which
exothermic

take in energy from the surroundings are called [8] _____ reactions.
endothermic

Bond making and breaking

| bond | bond breaking | bond making |

In chemical substances there are [H9] _____ between atoms.
bonds

In a chemical reaction there is a change in bonding. Energy is required for [H10] _____
bond

_____ and is released on [H11] _____ _____ .
breaking   *bond*   *making*

In an exothermic reaction less energy is required for [H12] _____ _____ than is
bond   *breaking*

released by [H13] _____ _____ . In an endothermic reaction less energy is
bond   *making*

required for [H14] _____ _____ than is released by [H15] _____
bond   *making*   *bond*

_____ .
breaking

Science: On Course for GCSE NEAB Edition Teacher's Book © Stanley Thornes (Publishers) Limited 1998

Energy level diagrams

activation energy	*catalyst*	*energy*	*energy change*	*energy level*
diagram	*exothermic*	*product*	*reactant*	

The minimum energy needed by reactant particles before a reaction can occur is called the

H16 _____ _____ .
     activation   *energy*

A substance which alters the rate of a reaction by lowering the activation energy is called a

H17 _____ .
     catalyst

Diagrams, like those below, that show energy changes during chemical reactions are called

H18 _____ _____ _____ . Label the diagrams using key words.
     energy   *level*   *diagrams*

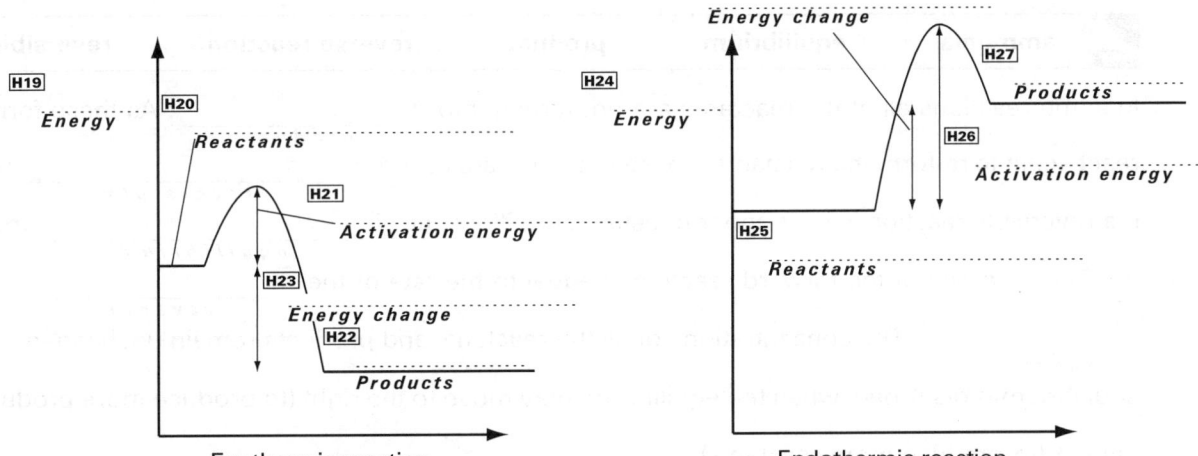

Exothermic reaction

Endothermic reaction

This diagram shows a test tube containing copper(II) sulphate solution. The thermometer shows the temperature of the solution. Some zinc powder is added to the test tube and the mixture is stirred. The new temperature is shown on the thermometer.

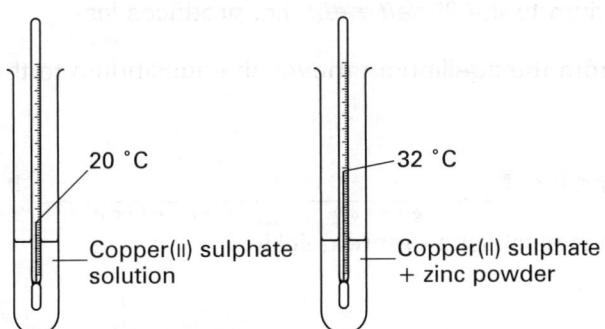

20 °C — Copper(II) sulphate solution

32 °C — Copper(II) sulphate + zinc powder

The reaction between copper(II) sulphate and zinc powder is an H29 _____
                                                                    exothermic

reaction. The reactants contain more energy than the H30 _____ .
                                                          products

Science: On Course for GCSE NEAB Edition Teacher's Book © Stanley Thornes (Publishers) Limited 1998

Chemicals from air

Do all of this topic for Double Science only.
H = for Higher Tier only

📝 Revision notes

Air is a raw material for producing many industrial chemicals.

Reversible reactions

🔑 **ammonia**   **equilibrium**   **product**   **reverse reaction**   **reversible**

In some reactions all of the reactants are not turned into [1] _____ *products* . As these form they

react again to re-form the reactants. These reactions are called [2] _____ *reversible* reactions.

If a reversible reaction is kept under constant conditions, an [3] _____ *equilibrium* may be set

up. Then the rate of the forward reaction is equal to the rate of the [4] _____ *reverse*

_____ *reaction* . The concentrations of all the reactants and products remain unchanged unless the

equilibrium is disturbed, when the equilibrium may move to the right (to produce more products) or to

the left (to produce more reactants).

An example of a reversible reaction is:

nitrogen + hydrogen $\rightleftharpoons$ [5] _____ *ammonia*

$$N_2(g) \;+\; 3H_2(g) \;\rightleftharpoons\; 2NH_3(g)$$

Decreasing the pressure moves the equilibrium to the [H6] *left*/~~right~~, i.e. produces less

[7] _____ *ammonia* . Removing ammonia from the equilibrium moves the equilibrium to the

[H8] ~~left~~/*right* producing more ammonia.

A catalyst speeds up the forward reaction and the [9] _____ *reverse* _____ *reaction* . It does not

produce more products but the equilibrium is established more quickly.

The Haber process

🔑 **catalyst**   **exothermic**   **hydrogen**   **iron**   **liquefying**
    nitrogen   **recycled**
    slow

Science: On Course for GCSE NEAB Edition Teacher's Book © Stanley Thornes (Publishers) Limited 1998

Ammonia gas is produced in the Haber process from three parts [10] _____ and one

_h y d r o g e n_

part [11] _____ by volume. Fractional distillation of liquid air produces

_n i t r o g e n_

[12] _____ and breaking down crude oil fractions or natural gas produces

_n i t r o g e n_

[13] _____. The gases are dried, mixed and compressed to a high pressure. They are then

_h y d r o g e n_

passed over a [14] _____ made of finely divided [15] _____ heated to about

_c a t a l y s t_ _i r o n_

450 °C. About 10% of the gases are converted to ammonia. The forward reaction is exothermic.

Decreasing the temperature will [H16] _increase_/~~decrease~~/~~not alter~~ the yield of ammonia. However it

will also [H17] _____ the rate of reaction. The ammonia is removed by

_s l o w_

[18] _____. The unreacted gases are [19] _____.

_l i q u e f y i n g_ _r e c y c l e d_

The catalyst does not have to be heated during the process because the reaction is

[20] _____.

_e x o t h e r m i c_

Nitric acid manufacture

oxygen	platinum	water

Nitric acid is produced by reacting ammonia with [21] _____ in the presence of a

_o x y g e n_

[22] _____ catalyst. Nitrogen monoxide is then cooled and reacted with oxygen

_p l a t i n u m_

and [23] _____ to produce nitric acid.

_w a t e r_

Fertilisers

ammonia	drinking water	nitric acid	nitrogen	sulphuric acid

The photograph shows a farmer spreading chemical fertiliser on the land.

The essential element provided by ammonium nitrate, ammonium sulphate and urea is

[24] _____. Ammonium nitrate is manufactured by the reaction of [25] _____ gas

_n i t r o g e n_ _a m m o n i a_

with [26] _____ _____. Ammonium sulphate is manufactured from ammonia

_n i t r i c_ _a c i d_

gas and [27] _____ _____. Ammonium nitrate and ammonium

_s u l p h u r i c_ _a c i d_

sulphate are very soluble in water and are therefore [28] _quick acting_/~~slow acting~~. Urea is almost

insoluble in water but reacts slowly to produce ammonia. It is therefore a [29] ~~quick acting~~/_slow acting_

fertiliser.

When nitrogen compounds are washed into rivers, they can cause problems when the water is used

as [30] _____ _____.

_d r i n k i n g_ _w a t e r_

Atomic structure and bonding

Do all of this topic for Double Science only.
H = for Higher Tier only

Revision notes

Matter made up of particles

boil	boiling point	condensing	density	diffusion	evaporating
freezing	gas	melting	random	vibrating	

All matter is made up of particles. In a solid the particles are usually close together and this leads to a solid having a high [1] _____ . The particles are [2] _____ . In a
                                density   *vibrating*
liquid, the particles are moving [3] *less*/more than in a solid. The arrangement of particles in a liquid is [4] *less regular*/more regular than in a solid.

In a gas the particles are [5] *close together*/widely spaced and moving rapidly in all directions. This is called [6] _____ motion.
            random
The movement of particles to fill all available space is called [7] _____ . This occurs
                                                    diffusion
most rapidly with a [8] _____ .
                    gas
The diagram below shows the changes of state between solids, liquids and gases. Label the diagram using key words.

[9] *Freezing* ____ / [10] ____ Solid

[11] *Melting* ____

Evaporating Liquid ———————— Gas

[12] ____
Condensing

If energy is given to a solid, its particles vibrate more. They may separate from each other and become free to move. This is called [13] _____ .
                                            melting
Heating a liquid makes its particles move around more quickly.

Particles which have enough energy may overcome attractive forces and escape from the liquid and become a [14] _____ . This is called [15] _____ .
            gas   *evaporating*

Science: On Course for GCSE NEAB Edition Teacher's Book © Stanley Thornes (Publishers) Limited 1998

At a higher temperature called the [16] _____ _____ the liquid will
boiling *point*

[17] _____ and become a gas.
boil

Chemical combination

| 🔑 | atom | combined | compound | magnet | mixture | sulphide | synthesis |

Elements are made up of tiny particles called [18] _____ . A piece of pure copper contains
atoms

only copper atoms.

A substance which contains atoms of different elements, and in which the atoms are not combined, is

called a [19] _____ . The atoms in a mixture can be separated. For example, iron can be
mixture

separated from a mixture of iron and sulphur using a [20] _____ . Atoms can be joined
magnet

together or [21] _____ to form a [22] _____ .
combined *compound*

A mixture of iron and sulphur, when heated, forms iron [23] _____ .
sulphide

The process of forming a compound from its constituent elements is called combination or

[24] _____ .
synthesis

Atomic structure

🔑	atomic number	chlorine	electron	energy	hydrogen
	isotope	mass number	neon	neutron	proton
	sodium				

Atoms are made up of three types of particles.

Use key words to complete this table:

particle	mass	charge
[25] _____ *neutron*	1	0
[26] _____ *electron*	negligible	−1
[27] _____ *proton*	1	+1

An atom consists of protons and neutrons tightly packed in the [28] _____ with
nucleus

[29] _____ moving rapidly around the nucleus in certain [30] _____
electrons *energy*

levels.

The nucleus of an atom is [31] *positively charged/~~negatively charged/neutral~~*.

A neutral atom contains equal numbers of [32] _____ *protons* in the nucleus and [33] _____ *electrons* outside the nucleus.

All oxygen atoms contain eight protons and eight [34] _____ *electrons* .

The only atom which does not contain one or more neutrons in the nucleus is the [35] _____ *hydrogen* atom, where the nucleus consists of a single [36] _____ *proton* .

When an ion is formed, an atom gains or loses [37] _____ *electrons* . If an atom loses one electron it forms a [38] ~~negatively charged~~/*positively charged* ion with a single charge.

Atoms of the same element with different mass numbers are called [39] _____ *isotopes* . Chlorine-35 and chlorine-37 are two [40] _____ *isotopes* of chlorine. An atom of chlorine-35 contains 17 protons, 17 [41] _____ *electrons* and 18 neutrons. Chlorine-35 and chlorine-37 have the same [42] _____ _____ *atomic numbers* but different [43] _____ _____ *mass numbers* .

The element with an electron arrangement of 2, 8, 1 is [44] _____ *sodium* .

The element with an electron arrangement of 2, 8, 7 is [45] _____ *chlorine* .

The element with two completely filled energy levels is [46] _____ *neon* .

Structure

| covalent | giant structure | ionic | iron | metallic | molecular |

Substances can be divided into those which have [H47] _____ *molecular* structures and those which have [H48] _____ _____ *giant structure* . The photograph shows sodium chloride crystals. A regular shaped crystal is evidence for [H49] *regular*/~~irregular~~ arrangements of particles.

Substances with giant structures have [H50] *high*/~~low~~ melting points and boiling points. Substances with molecular structures have [H51] ~~high~~/*low* melting points and boiling points.

There are three types of forces present between particles. They are:

[H52] *m* _____ *metallic* , [H53] *c* _____ *covalent* and [H54] *i* _____ *ionic* .

Complete this table by using key words:

example	type of structure	type of bonding	solubility in water	conductivity of molten substance
[H55] ___ *iron*	[H56] ___ *giant* ___ *structure*	metallic	does not dissolve	conducts when solid and molten
sodium chloride	[H57] ___ *giant* ___ *structure*	[H58] ___ *ionic*	soluble	conducts
boron oxide	[H59] ___ *giant* ___ *structure*	[H60] ___ *covalent*	does not dissolve	does not conduct

Ionic and covalent bonding

chlorine	electrostatic	ion	lattice	sodium
bonding	*covalent*	*cross-link*	*electron*	*molecule*
pair	*thermosetting*	*thermosoftening*		

The forces which hold atoms together are called [H61] _____ .
bonding

This diagram shows the arrangement of electrons in sodium and chlorine atoms.

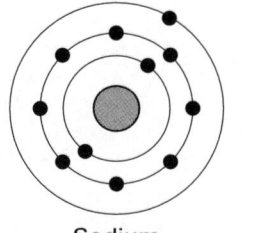 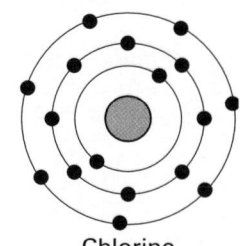

Sodium   Chlorine

When sodium and chlorine combine, one electron is transferred from the [62] _____ atom
sodium

to the [63] _____ atom. Sodium and chloride [64] _____ are held together in a
chlorine   *ions*

structure or [65] _____ by strong [66] _____ forces.
lattice   *electrostatic*

The diagram on the left shows the arrangement of electrons in a carbon atom and four hydrogen

atoms.

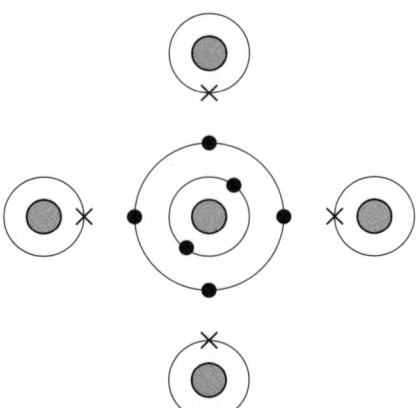 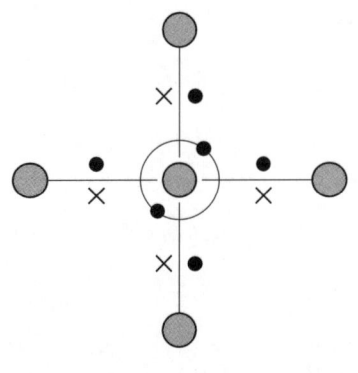

When the carbon atom and the four hydrogen atoms combine they form a [H67] _____ of
molecule

methane, CH_4. Each bond uses one electron from the carbon atom and one electron from a hydrogen

atom to form a shared [H68] _____ of electrons. Within each methane molecule there are four
pair

[H69] _____ bonds. The forces between methane molecules are [H70] ~~very strong~~/very weak.
covalent

The diagram on the right above shows a methane molecule.

An oxygen molecule contains a double [H71] _____ bond. In this bond there are four
covalent

shared [H72] _____ .
electrons

Plastics are polymers with a tangled mass of very long molecules. A plastic consisting of chains with only weak forces between the chains will melt on heating. Plastics of this type are called [H73] _____ *thermosoftening* plastics. Plastics which have covalent [H74] _____ *cross-links* between the chains will not melt easily. These are called [H75] _____ *thermosetting* plastics.

The Periodic Table

Do all of this topic for Single Science and Double Science.
H = for Higher Tier only

Revision notes

Properties of elements

acidic	alkaline	chlorine	helium	inert	metal

An element which can be used to fill the balloon in the photograph is [1] _____ . Unlike
helium

hydrogen, this gas is very unreactive or [2] _____ .
inert

An element used to make household bleaches to kill germs is [3] _____ .
chlorine

An element which is shiny, conducts heat and electricity and has a high density is a

[4] _____ .
metal

Metals burn in oxygen to form oxides which are neutral or [5] _____ . Non-metals form
alkaline

oxides which are neutral or [6] _____ .
acidic

Arrangement of elements in the Periodic Table

alkali metal	atomic mass	atomic number	group
halogen	metal	noble gas	non-metal
period	transition metal	trend	

A horizontal row in the Periodic Table is called a [7] _____ and a vertical column is called
period

a [8] _____ . The elements in Mendeleyev's original Periodic Table were arranged in order
group

of increasing [9] _____ _____ but in the modern Periodic Table they are
atomic *mass*

arranged in order of increasing [10] _____ _____ .
atomic *number*

In any period of the Periodic Table, the elements change across the period from [11] _____
metals

on the left-hand side to [12] _____ on the right-hand side. Elements in the same
non-metals

group are similar but not identical. There are patterns or [13] _____ within each group. In
trends

group 4 carbon is a non-metal, silicon is a non-metal with metallic appearance, and tin and lead are

[14] _____ .
metals

Science: On Course for GCSE NEAB Edition Teacher's Book © Stanley Thornes (Publishers) Limited 1998

Elements in group 1 of the Periodic Table are called [15] _____ _____ and
alkali *metals*

elements in group 7 are called [16] _____ . The unreactive elements in group 0 are
halogens

called [17] _____ _____ .
noble *gases*

The block of elements between groups 2 and 3 in the Periodic Table is called the

[18] _____ _____ . Transition metals are less reactive than alkali metals.
transition *metals*

Transition metal compounds are often coloured, e.g. iron(II) sulphate crystals are

[19] *blue*/green/*brown*/*red*.

Lithium

🔑	alkali	hydrogen	oxide	paraffin oil	shiny

Lithium is an alkali metal. It is stored under [20] _____ _____ because it is so
paraffin *oil*

reactive.

When a piece of lithium is cut with a knife it leaves a [21] _____ surface which soon goes
shiny

dull owing to the formation of lithium [22] _____ .
oxide

Lithium reacts with water to form lithium hydroxide and [23] _____ .
hydrogen

When lithium hydroxide is tested with universal indicator, it turns purple, showing that lithium

hydroxide is an [24] _____ .
alkali

Bromine

🔑	chlorine	iodine	liquid	salt

Bromine is an element in group 7 of the Periodic Table and it is a [25] _____ non-metal at
liquid

room temperature. Bromine will react with a metal to form a [26] _____ . It is more reactive
salt

than [27] _____ but less reactive than [28] _____ . Bromine will displace
iodine *chlorine*

[29] _____ from potassium iodide solution but not [30] _____ from potassium
iodine *chlorine*

chloride solution.

Chlorides

🔑	chlorine	hydrogen chloride	silver chloride	sodium chloride
	sodium hydroxide			

When silver nitrate solution is added to a solution of sodium chloride, a white precipitate of

[31] _____ _____ is formed.
silver *chloride*

Hydrochloric acid is made by dissolving [32] _____ _____ gas in water.

hydrogen *chloride*

Brine is a solution of [33] _____ _____ in water. Electrolysis of brine produces

sodium *chloride*

hydrogen gas, [34] _____ gas and [35] _____ _____ solution.

chlorine *sodium* *hydroxide*

Skeleton Periodic Table

alkali metal	group	halogen	noble gas	period
energy level				

Elements in the same group each contain the same number of electrons in the outer

[H36] _____ _____ . The higher the energy level the more easily electrons

energy *level*

are [H37] ~~gained~~/lost and the less easily electrons are [H38] gained/~~lost~~ .

Here is a skeleton Periodic Table. Some of the elements are represented by the letters **A, B, C** and **D**.

The element shown by the letter **B** is in the [39] _____ _____ family.

alkali *metal*

The element shown by the letter **D** is in the [40] _____ family.

halogen

The element shown by the letter **C** is in the [41] _____ _____ family.

noble *gas*

Elements in this group have a full outer [H42] _____ _____ and no tendency to

energy *level*

gain or lose or share [H43] _____ .

electrons

Elements **B** and **D** are in the same [44] _____ of the Periodic Table.

period

Elements **A** and **B** are in the same [45] _____ of the Periodic Table.

group

Uses

acidic	chlorine	sodium chloride	sodium hydroxide

Here is a list of uses of sodium compounds. From the list of key words select one compound which is

suitable for each use.

Flavouring and preserving food   [46] _____ _____

sodium *chloride*

Reacting with natural fats to make soap   [47] _____ _____

sodium *hydroxide*

Science: On Course for GCSE NEAB Edition Teacher's Book © Stanley Thornes (Publishers) Limited 1998

Household bleaches can be produced by reacting [48]_____ gas with
chlorine
[49]_____ _____ solution.
sodium   *hydroxide*
Hydrogen halides are gases which dissolve in water to form [50]_____ solutions.
acidic

Science: On Course for GCSE NEAB Edition Teacher's Book © Stanley Thornes (Publishers) Limited 1998

Transferring energy

Do all of this topic for Single Science and Double Science.
H = for Higher Tier only

 ## Revision notes

Methods of energy transfer

absorber	convection	density	electromagnetic radiation	emitter
gas	infra-red	insulator	metal	particle
radiation	temperature	thermal energy		
diffusion	*expands*	*free electron*	*kinetic*	

There is a flow of energy between places at different [1] _____ . This flow can take
temperatures

place in three main ways.

These processes are conduction, [2] *c*_____ and [3] *r*_____ . Of
convection   *radiation*

these three, only [4] _____ can transfer energy through a vacuum.
radiation

Heat ([5] _____ _____) can pass through all materials by conduction.
thermal   *energy*

Conduction relies on the movement of [6] _____ ; when particles become more
particles

energetic they transfer energy through collisions with neighbouring particles. The best conductors

are [7] _____ and the worst are [8] _____ . Poor conductors are called
metals   *gases*

[9] _____ .
insulators

Metals are much better conductors than non-metals because they have [H10] _____
free

_____ that can move within the metal. Heating one part of a metal gives these
electrons

particles more [H11] _____ energy which is transferred to other parts of the metal by
kinetic

[H12] _____ .
diffusion

Convection involves a flow of material, so it only takes place in liquids and [13] _____ .
gases

Convection currents are caused by changes in [14] _____ . When the air around a central
density

heating radiator is heated it [H15] _____ , becoming less dense. The warmed air rises above
expands

the surrounding colder, denser air.

Science: On Course for GCSE NEAB Edition Teacher's Book © Stanley Thornes (Publishers) Limited 1998

All objects emit energy in the form of [16] _____ _____ .
electromagnetic *radiation*

These waves are in the [17] _____ region of the electromagnetic spectrum. The
infra-red

hotter the object, the more energy it radiates each second. Very hot objects also emit light and other

electromagnetic waves.

Some objects are better than others at absorbing and emitting radiant energy. Dark colours are good

absorbers and good [18] _____ of infra-red radiation. Light colours are poor emitters and
emitters

poor [19] _____ of infra-red radiation. Silvered surfaces
absorbers

reflect [20] _____ radiation in the same way that they reflect light.
infra-red

Insulation

	conduction	conductor	convection	insulation	insulator
	loft insulation	reflects			

Keeping things warm in a cold environment requires [21] _____ . If a house is
insulation

warmer than its surroundings, it loses most energy by conduction and [22] _____ .
convection

The diagram shows the energy flow through an uninsulated cavity wall. The energy flows through the

brick by [23] _____ and through the air-filled cavity by
conduction

[24] _____ .
convection

Cold outside   Warm inside

The rate of energy flow through the wall can be reduced by installing cavity wall insulation. This uses

foam or mineral wool to trap pockets of air. If the air cannot move then [25] _____
convection

currents cannot flow. Energy can still flow by the process of [26] _____ but gases
conduction

are very poor [27] _____ .
conductors

Other methods of insulating a house include loft insulation and fitting double glazing. Of these,

[28] _____ _____ is the most cost-effective.
loft *insulation*

A hot drink or food taken from a hot oven is a lot warmer than its surroundings. It loses energy mainly by infra-red radiation and evaporation. Aluminium foil is a very good [29] _____ for hot
insulator

food and drink because it [30] _____ infra-red radiation and prevents hot vapour from escaping.
reflects

Science: On Course for GCSE NEAB Edition Teacher's Book © Stanley Thornes (Publishers) Limited 1998

Generating and using electricity

Do all of this topic for Single Science and Double Science.

📝 Revision notes

Energy from electricity

🔑	electricity	efficiency	energy	heat	kilowatt-hour	light
	movement	power	sound	watt		

The most convenient source of [1] _____ for use at home and at work is electricity. Energy

energy

from electricity is easily transferred as [2] h_____ , [3] m_____ , [4] l_____

heat *movement* *light*

and [5] s_____ (a form of movement). A hairdryer is designed to transfer energy from

sound

electricity as [6] h_____ and [7] m_____ of the air. It also transfers some energy

heat *movement*

as [8] _____ . A television is designed to produce [9] l_____ and

sound *light*

[10] s_____ but it also produces [11] _____ .

sound *heat*

Kettles and immersion heaters are designed to transfer energy from electricity to

[12] _____ in the water. They are very efficient at doing this. Filament lamps only transfer a

heat

small amount of the energy from [13] _____ as light; they have a low

electricity

[14] _____ . Energy-efficient lamps produce the same [15] _____ output

efficiency *light*

as filament lamps but take in less energy from [16] _____ .

electricity

Electrical power is measured in watts (W) or kilowatts (kW) and is calculated from the equation

$$power = \frac{energy\ transfer}{time\ taken}$$

One kW is equal to 1000 W. High [17] _____ appliances such as kettles and immersion

power

heaters transfer energy at a greater rate than low power appliances such as lamps. Electricity supply

companies measure the amount of energy transfer from the mains in [18] _____

kilowatt-hours

(kWh) which can be calculated using the equation

energy transfer in kWh = power in kW × time in h

Science: On Course for GCSE NEAB Edition Teacher's Book © Stanley Thornes (Publishers) Limited 1998

The same equation is used to calculate an energy transfer in joules, using the power in

[19] _____ and the time in seconds.
watts

Fossil fuels

🔑	coal	electricity	fossil fuel	generator	non-renewable	oil	pressure
	Sun	temperature	turbine				

Most of our electricity is generated by burning [20] _____ _____ such as coal,
fossil   *fuels*

gas and [21] _____ . No more of these fuels can be made; they are
oil

[22] _____ . Like most of our energy sources, the energy stored in fossil fuels came
non-renewable

from the [23] _____ . The energy was stored by plants through photosynthesis.
Sun

In a coal-fired power station energy obtained from burning the [24] _____ is used to
coal

generate steam at very high temperature and [25] _____ . The steam transfers its energy as
pressure

it passes through the [26] _____ that drive the [27] _____ , producing
turbines   *generator*

electricity.

The maximum efficiency of a coal-fired power station is 0.45 or 45%. This means that 45%, or less

than half, of the energy from the fuel is transferred to [28] _____ . The remainder
electricity

goes into the surroundings. This energy is very difficult to recover because it causes only a small rise

in [29] _____ .
temperature

Nuclear power and geothermal energy

🔑	electricity	geothermal	radioactive	turbine

Steam-driven turbines are also used in nuclear power stations to generate electricity. Nuclear power

stations produce [30] _____ waste that cannot be disposed of easily. They are also
radioactive

expensive to build and to dispose of when they reach the end of their useful lives.

Energy from underground rocks that are heated by [31] _____ decay is known
radioactive

as [32] _____ energy. Cold water can be pumped into the rocks and returned as
geothermal

hot water to be used for heating. If the rocks are hot enough this can be used to generate

[33] _____ using a steam-driven [34] _____ .
electricity   *turbine*

Science: On Course for GCSE NEAB Edition Teacher's Book © Stanley Thornes (Publishers) Limited 1998

Renewable energy sources

atmosphere	battery	efficiency	electricity	generator
gravitational potential	hydroelectric	kinetic	mains electricity	noise
power	renewable	Sun	turbine	wave

The Sun, wood, wind and moving water are all examples of [35] _____ energy
renewable
sources. Energy in the wind, [36] _____ and rivers comes from the Sun heating the
waves
atmosphere; it can be used to drive [37] _____ directly without using steam.
generators
Wind turbines and [38] _____ power stations have the benefit of not polluting
hydroelectric
the [39] _____ , but they are expensive to build. A wind farm takes up a large
atmosphere
amount of space and also causes [40] _____ pollution.
noise
Wave power is another [41] _____ energy source, but it is proving difficult to
renewable
obtain energy from the waves in a cost-effective way.

Solar cells, which produce [42] _____ from the Sun's radiation, are also expensive
electricity
and take up a large area. They can be used on the walls of large buildings to provide some electricity
and, together with storage [43] _____ , they are useful for providing small amounts
batteries
of electricity in remote places where there is no [44] _____ _____ .
mains   *electricity*
They are also useful for low [45] _____ devices such as calculators.
power
Solar heating uses energy from the [46] _____ directly to heat water. This can be done with
Sun
high [47] _____ by passing water through blackened pipes that absorb radiant
efficiency
energy from the Sun.

Energy can be usefully stored as gravitational potential energy (gpe), e.g. in a pumped storage

scheme. There is low demand for electricity at night, so it is used to pump water from a low reservoir

to a high one. At times of peak demand the water is released. As it falls downhill, it loses

[48] _____ _____ energy and gains [49] _____
gravitational   *potential*   *kinetic*
energy which is then transferred to electricity as the water passes through the [50] _____ .
turbines

Water released to
generate electricity
at peak demand

Water pumped to high
reservoir at night

Turbines and
pumps/generators

The change in gravitational potential

energy when water moves up or

downhill is calculated using the equation

change in gpe = weight × change in height

Science: On Course for GCSE NEAB Edition Teacher's Book © Stanley Thornes (Publishers) Limited 1998

Current, charge and circuits

Do part of this topic for Single Science and all of it for Double Science.
D = for Double Science only   H = for Higher Tier only

📝 Revision notes

Charged particles

🔑 | **attract** | **electron** | **force** | **repel** |

Charged objects exert [D1] _____ on each other. Objects with a similar
forces

charge [D2] _____ each other and those with opposite charges [D3] _____ each
repel   *attract*

other. Objects become charged by adding or removing [D4] _____ .
electrons

Static charge

🔑 | **conductor** | **friction** | **negatively** | **positively** | **static** |
| *voltage* | | | | |

The [D5] _____ forces that exist when two objects rub against each other cause the transfer
friction

of electrons. The object that gains electrons becomes charged [D6] _____ while
negatively

the object that loses electrons becomes charged [D7] _____ . Insulators keep this
positively

charge but conductors quickly lose it as it passes through them and to earth.

Charge that is not moving is said to be [D8] _____ . It can be both useful and dangerous. A
static

build up of static charge can create a high [DH9] _____ that can cause lightning or sparks,
voltage

creating a fire hazard. When an aircraft is being refueled, sparking is avoided by connecting the aircraft

to earth using a good electrical [D10] _____ . Photocopiers use
conductor

[D11] _____ charge to attract the black powder to the paper and coal-burning power stations
static

use it to remove dust from the waste gases.

Electric current

🔑 | **ammeter** | **amp** | **current** | **dissolved** |
| **electron** | **negative** | **positive** | |
| *current* | *time* | | |

Science: On Course for GCSE NEAB Edition Teacher's Book © Stanley Thornes (Publishers) Limited 1998

A flow of charge is called an electric [12] _current_ . Current is measured in

[13] _amps_ using an [14] _ammeter_ . Current in a metal is due to a flow of

[D15] _electrons_ that move away from the [D16] _negative_ terminal of the battery

or power supply and towards the [D17] _positive_ terminal. Current in conducting gases and

molten or [D18] _dissolved_ ionic compounds is due to the movement of both positive

and [D19] _negative_ ions.

During electrolysis negative ions are attracted to the [D20] _positive_ terminal and positive ions

are attracted to the [D21] _negative_ terminal. This causes the deposit of solids and the release

of gases at the terminals. The amount of substance released is increased by increasing the

[DH22] _current_ or the [DH23] _time_ for which the current passes.

Circuits

ammeter	battery	closed switch	component	current
diode	lamp	parallel	resistance	resistor
series	variable resistor	voltage	voltmeter	

A circuit that has only one path for the current is a [24] _series_ circuit. Where there are two or

more possible current paths the circuit is a [25] _parallel_ circuit. The current is the same at all

points in a series circuit; no charge is gained or lost. The total [26] _resistance_ in a series

circuit is equal to the sum of the resistances of the components. The supply [27] _voltage_ is

shared between the components.

Components connected in parallel have the same [28] _voltage_ across them. The greater the

resistance of a component, the smaller the [29] _current_ . The total current is the sum of the

currents in the individual [30] _components_ .

Here is a list of circuit symbols for some common devices. Use key words to complete the labels.

Cell

[31] _Battery_

Open switch

[32] _Closed switch_

[33] _Diode_

Light-dependent resistor

Thermistor

[34] _Resistor_

[35] _Variable resistor_

[36] (A) _Ammeter_

[37] (V) _Voltmeter_

[38] _Lamp_

Voltage

🔑	energy	parallel	voltage	voltmeter

The job of the current in a circuit is to transfer **[39]** _____ from the power supply to the
e n e r g y

circuit components. Energy is transferred as heat, light and movement in the circuit components.

Voltage is measured using a **[40]** _____ that is placed in **[41]** _____ with
v o l t m e t e r   *p a r a l l e l*

a power supply or electrical device.

The amount of current passing in a circuit depends on the **[42]** _____ and the resistance of
v o l t a g e

the circuit. Increasing the voltage causes the current to **[43]** *decrease/stay the same/increase*, while

increasing the resistance causes the current to **[44]** *decrease/stay the same/increase*.

Resistance

🔑	current	ohm	voltmeter

The voltage across a circuit component is equal to the current times its resistance, $V = I \times R$.

This equation is also used in the form *resistance = voltage/current*, $R = V/I$ to calculate resistance

from ammeter and **[H45]** _____ readings. Resistance is measured in
v o l t m e t e r

[H46] _____ .
o h m s

Provided that the temperature stays the same, the resistance of a metal wire does not change when

the current changes. A graph of current against voltage is a straight line passing through the origin.

The wire in a filament lamp gets hotter as the current increases and this causes the resistance to

[H47] *decrease/stay the same/increase*.

A diode only allows **[H48]** _____ to pass in one direction. The resistance of a diode that is
c u r r e n t

conducting **[H49]** *decreases/increases* when the current is increased.

The resistance of a light-dependent resistor (LDR) and a thermistor depend on environmental

conditions; that of an LDR decreases as the light level increases and that of a

thermistor **[H50]** *decreases/increases* as the temperature increases.

Science: On Course for GCSE NEAB Edition Teacher's Book © Stanley Thornes (Publishers) Limited 1998

Using electricity

Do part of this topic for Single Science and all of it for Double Science.
D = for Double Science only   H = for Higher Tier only

Revision notes

Electricity in the home

alternating	circuit breaker	conductor	current	direct
earth	fuse	insulated	insulation	live
neutral	resistance			

The current due to a cell or battery in a circuit is always in the same direction; it is called a

[1] _____ current (d.c.). A current that changes direction is called an
direct

[2] _____ current (a.c.). Mains electricity uses alternating current that changes
alternating

direction 100 times each second.

The three conductors that form the electricity cable to a house are called the [3] / _____ ,
live

[4] n _____ and [5] e _____ . The [6] _____ wire alternates between a
neutral   *earth*   *live*

positive and negative voltage relative to the neutral. The [7] _____ wire is the return path
neutral

that completes the circuit and the [8] _____ wire is for safety. When an appliance is
earth

operating normally there is no current in the [9] _____ wire.
earth

A metal-cased appliance such as a toaster needs all three conductors. The flexible cable that connects

the toaster to the mains supply has three separate wires, each of which has a layer of

[10] _____ .
insulation

The wire that has blue insulation is the [11] _____ , the wire with brown insulation is the
neutral

[12] _____ and that with green and yellow insulation is the [13] _____ .
live   *earth*

At the toaster, the live and [14] _____ are connected to the heating element and
neutral

the [15] _____ is connected to the metal case. The switch and a [16] _____ are
earth   *fuse*

also connected in the live conductor as shown in the diagram.

Science: On Course for GCSE NEAB Edition Teacher's Book © Stanley Thornes (Publishers) Limited 1998

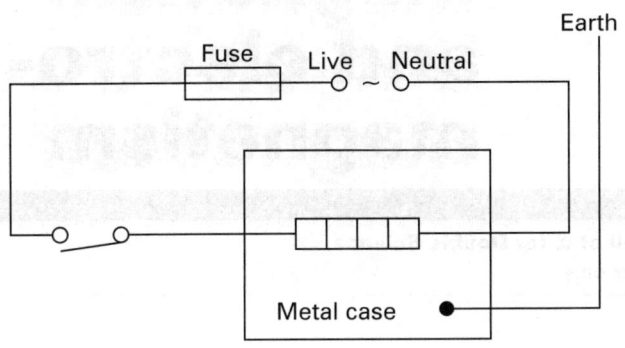

The $\boxed{17}$ _____ protects against a fire hazard in the connecting wires. If a fault in the
       f u s e

element causes too large a $\boxed{18}$ _____ the fuse melts and breaks the circuit. The fuse and
              c u r r e n t

the $\boxed{19}$ _____ wire together protect the user from electrocution. If the case becomes live
       e a r t h

the earth wire provides a low $\boxed{20}$ _____ path to earth. This causes a large current
                r e s i s t a n c e

and the $\boxed{21}$ _____ melts, breaking the circuit.
        f u s e

Appliances such as televisions and hairdryers normally have plastic cases which cannot become live

because they are not $\boxed{22}$ _____ . These appliances are said to be
               c o n d u c t o r s

double $\boxed{23}$ _____ and they do not need an $\boxed{24}$ _____ wire.
        i n s u l a t e d   *e a r t h*

In addition to the fuse fitted to the plug of each appliance, each circuit in a house has its own fuse or

$\boxed{25}$ _____ _____ to protect the fixed cables from overheating and causing a
    c i r c u i t   *b r e a k e r*

fire. Circuit breakers are more reliable than $\boxed{26}$ _____ and they are easily reset when the
                 f u s e s

fault has been put right.

Energy transfer from electricity

energy	voltage	watt
coulomb		

All electrical appliances have a power rating. This is the rate at which $\boxed{27}$ _____ is
                                 e n e r g y

transferred from the electricity supply. Power is measured in $\boxed{28}$ _____ or kilowatts and
                                w a t t s

can be calculated using the equation *power = current* × $\boxed{\text{D29}}$ _____ or $P = I \times V$.
                                 v o l t a g e

The unit of charge is the $\boxed{\text{DH30}}$ _____ ; the amount of charge that flows when a current
                   c o u l o m b

passes can be calculated using the equation *charge = current* × *time*.

Voltage is a measure of energy transfer (in joules) for each coulomb of charge. The energy transfer by

charge flowing can be calculated using *energy = voltage* × *charge*.

Magnetism and electro-magnetism

Do part of this topic for Single Science and all of it for Double Science.
D = for Double Science only   H = for Higher Tier only

Revision notes

Magnetic poles

| attract | magnetic field | north | pole | repel | south |

Magnets attract objects made out of magnetic materials such as iron, steel and nickel. They can

D1 *a* attract_____ and **D2** *r* repel_____ other magnets. The strongest parts of a magnet are

called the **D3** _____ poles . A fixed magnet has two poles, called the north and south poles. The

north (or north-seeking) pole of a magnet is attracted to the **D4** _____ North pole of the Earth

and the south (or south-seeking) pole is attracted to the Earth's **D5** _____ South pole. Similar

magnetic poles **D6** _____ repel each other and opposite poles **D7** _____ attract .

Any region where a force acts on magnetic materials is called a **8** _____ magnetic

_____ field .

Electromagnets

| bar magnet | coil | current | electromagnet |

The most useful magnets are those that can be switched on and off; these are called

D9 _____ electromagnets .

Every electric **D10** _____ current has its own magnetic field. A current passing in a

D11 _____ coil of wire creates a magnetic field both inside and around the coil similar to that of

a **D12** _____ bar _____ magnet . Using an **D13** ~~brass/plastic~~/iron core creates a much

stronger electromagnet. The core is quickly magnetised when the current passes in the coil and it

loses its magnetism quickly when the current is switched off.

Science: On Course for GCSE NEAB Edition Teacher's Book © Stanley Thornes (Publishers) Limited 1998

Loudspeaker and relay

alternating	armature	attracted	coil	current
fixed magnet	frequency	iron core	paper cone	switch contact

The diagrams show two devices that use electromagnets, a loudspeaker and a relay.

Complete the labels by using key words.

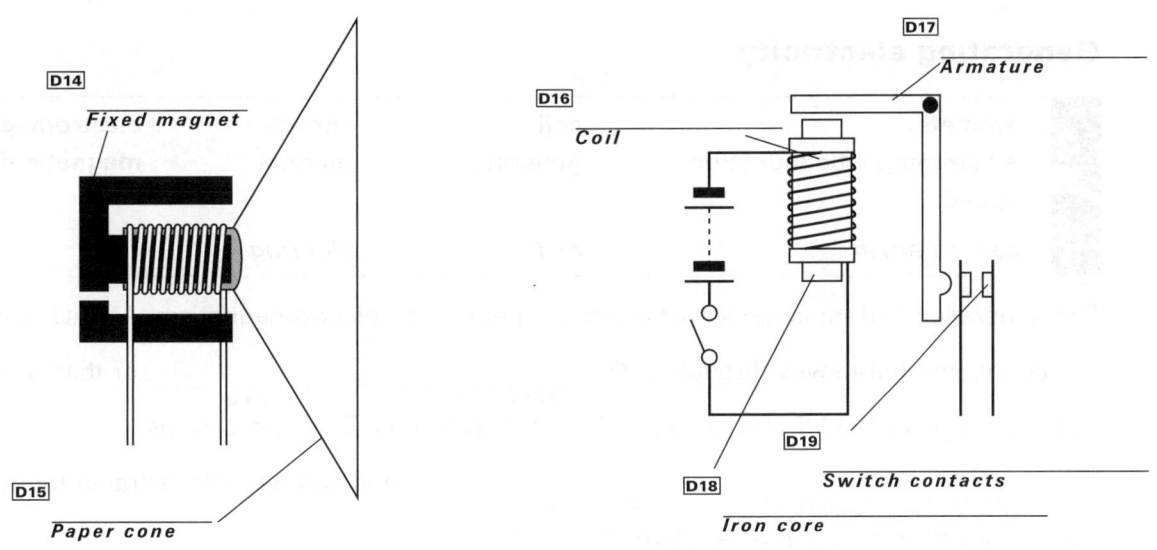

D14 | *Fixed magnet*

D15 | *Paper cone*

D16 | *Coil*

D17 | *Armature*

D18 | *Iron core*

D19 | *Switch contacts*

When a current passes in the D20 _____ of the loudspeaker, the paper cone is pushed in or
coil

out, depending on the direction of the D21 _____. An D22 _____
current   *alternating*

current causes it to move in and out repeatedly at the same D23 _____ as the
frequency

current.

The relay is a device that enables a small D24 _____ to switch a much larger current on
current

and off. When a small current passes in the coil, the magnetic field magnetises the iron core. This

causes the D25 _____ to be D26 _____ to the iron core, pressing the
armature   *attracted*

switch contacts together.

Electric motors

electromagnetism	magnetic field

Electric motors also rely on D27 _____. They use the principle that when an
electromagnetism

electric current passes in a wire placed at right angles to a D28 _____ _____,
magnetic   *field*

there is a force on the wire.

Science: On Course for GCSE NEAB Edition Teacher's Book © Stanley Thornes (Publishers) Limited 1998

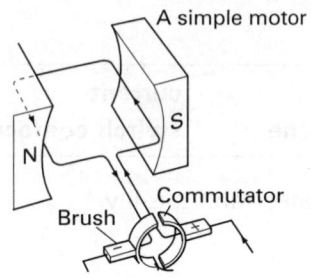

A simple motor

Brush

Commutator

N

S

In a motor, a coil of wire is placed between two opposite magnetic poles. The forces on the sides of the coil are in [D29] *opposite directions*/~~the same direction~~.

Generating electricity

ammeter	coil	current	electromagnet
electromagnetic induction	generator	magnet	magnetic field
speed			
carbon brush	*coil*	*slip ring*	

The generation and transmission of electricity depend on electromagnetism. A voltage is created in any conductor that moves through a [30] _____*magnetic*_____ _____*field*_____ or that is positioned within a magnetic field that is changing in size or direction. This is known as

[31] _____*electromagnetic*_____ _____*induction*_____ and can be demonstrated by using a magnet, a coil of wire and a sensitive [32] _____*ammeter*_____ .

The ammeter detects a [33] _____*current*_____ whenever the coil or the [34] _____*magnet*_____ is moved. The size and direction of the current depend on the

[35] _____*speed*_____ and direction of movement.

N S

Sensitive ammeter

A bicycle dynamo generates electricity by rotating a [36] _____*magnet*_____ next to a coil of wire. A power station generator works in a similar way; in this case an [37] _____*electromagnet*_____ is rotated inside the thick wire coils.

The diagram shows an a.c. [38] _____*generator*_____ .

Coil

Axle

N

S

Slip rings

Alternating voltage

Carbon brushes

The size of the induced voltage can be increased by increasing the [39] _____ of the coil,
speed

the strength of the [40] _____ _____ , the number of turns on the
magnetic   *field*

[41] _____ or the area of the [H42] _____ .
coil   *coil*

The [H43] _____ _____ provide a fixed contact to the rotating coil and the
slip   *rings*

[H44] _____ _____ provide a path for the current to pass out of the coil.
carbon   *brushes*

Transformers

current	voltage
primary	*secondary*

Transformers are used in the electricity supply industry. To minimise the energy wasted as heat in the

transmission wires, electricity is distributed at a high [DH45] _____ to keep the
voltage

[DH46] _____ as low as possible. Step-up transformers are used at power stations to
current

increase the [D47] _____ before the electricity is fed into the grid.
voltage

The [D48] _____ is stepped down in stages before being supplied to homes and
voltage

workplaces.

The formula that relates the primary and secondary voltages to the number of turns on the coils is:

$$\frac{[DH49] \underline{\hspace{2cm}} \text{ voltage}}{[DH50] \underline{\hspace{2cm}} \text{ voltage}} = \frac{\text{number of primary turns}}{\text{number of secondary turns}}$$

primary

secondary

Science: On Course for GCSE NEAB Edition Teacher's Book © Stanley Thornes (Publishers) Limited 1998

Force and motion

Do part of this topic for Single Science and all of it for Double Science.
D = for Double Science only   **H** = for Higher Tier only

📝 Revision notes

Graphs of motion

🔑 | accelerating   acceleration   distance   gradient   speed   velocity
acceleration   *distance*

The average speed of a moving object is calculated using the formula:

$$\boxed{1}\underline{\quad\quad\quad\quad} = \frac{distance\ travelled}{time\ taken}$$
speed

A distance–time graph shows the total distance travelled by an object at each point of its motion. The

slope or ☐2 _____ of the graph at any point represents the object's speed.
gradient

If you walk to the shop and then return home, the ☐3 _____ you have travelled
distance

increases throughout the journey. The graph shows how the distance travelled changes on such a

journey.

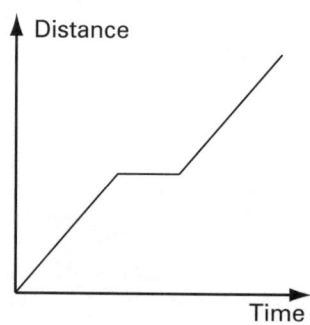

The gradient of a distance–time graph can only be positive; the steeper the gradient, the faster the

☐4 _____ it represents.
speed

Velocity is the speed in a given direction. The gradient of a velocity-time graph represents

acceleration. A positive gradient shows an increase in velocity and a negative gradient shows a

decrease in ☐D5 _____ .
velocity

Science: On Course for GCSE NEAB Edition Teacher's Book © Stanley Thornes (Publishers) Limited 1998

Electromagnetic radiation

Do part of this topic for Single Science and all of it for Double Science.
D = for Double Science only   **H** = for Higher Tier only

 Revision notes

Light

critical angle	dispersion	mirror	reflection
refraction	sound	spectrum	total internal reflection
wavelength			

Light travels as a transverse wave that has a much shorter [1] _____ and travels
wavelength

faster than [2] _____ . It is reflected in all directions by rough surfaces and in a predictable
sound

way by shiny surfaces and [3] _____ , the angle of incidence and the angle of
mirrors

[4] _____ being equal.
reflection

Reflection of light also takes place at the internal surface of glass and perspex. Provided that the angle

of incidence is greater than the [D5] _____ _____ , all the light is reflected when
critical   *angle*

it hits the boundary. This is called [D6] _____ _____
total   *internal*

_____ and is used in reflecting prisms and in optical fibres that transmit
reflection

information.

Light changes speed and [7] _____ when it passes from one substance to another.
wavelength

This is called [8] _____ and it can cause a change in direction.
refraction

When light is refracted, different colours change speed by different amounts. This leads to the

[9] _____ of light, which is the spreading of the colours into a
dispersion

[10] _____ . It is particularly noticeable when light passes through a triangular prism.
spectrum

Science: On Course for GCSE NEAB Edition Teacher's Book © Stanley Thornes (Publishers) Limited 1998

Light has a very short [11] _____ wavelength , typically 0.5 μm (1 μm = 1 millionth of a metre),
and so diffraction effects are difficult to observe.

Electromagnetic spectrum

gamma ray	**electromagnetic spectrum**	**frequency**	**infra-red**
light	**microwave**	**prism**	**radio wave**
total internal reflection	**ultraviolet**	**vacuum**	**wavelength**
diffracted			

Light is only a small part of the [12] _____ electromagnetic _____ spectrum , a family of waves
that all travel at the same speed in a [13] _____ vacuum .

The electromagnetic spectrum is a family of waves of the same type that differ in wavelength and

[14] _____ frequency . It extends from [15] _____ radio _____ waves , which have

the longest [16] _____ wavelength and lowest [17] _____ frequency , to X-rays and

[18] _____ gamma _____ rays , which have a very high [19] _____ frequency and

short [20] _____ wavelength .

The diagram shows where each electromagnetic wave fits into the spectrum.

```
Frequency (Hz)

     10²⁰        10¹⁷          10¹⁴          10¹¹          10⁸           10⁵

 ──────────────▶                ◀──────▶                ◀──────▶
  Gamma rays              Ultraviolet            Microwaves

 ──────────────▶        ◀───────────────────────▶              ◀──────────────
       X-rays                    Infra-red                         Radio waves

                            ⚹
                          Light

  10⁻¹²        10⁻⁹          10⁻⁶          10⁻³           1            10³
  Wavelength/(m)
```

Radio and television programmes are broadcast using [21] _____ radio _____ waves .

Long wavelength radio waves follow the Earth's curvature and are readily

[H22] _____ diffracted around buildings and hills.

[23] _____ Microwaves are used both for cooking and for radio transmissions. Television

remote controls and oven grills use [24] _____ infra-red . Our eyes detect

[25] _____ light , which can travel down optical fibres by repeated [26] _____ total

_____ internal _____ reflection .

Total internal reflection is also used by reflecting [27] _____ in cameras and binoculars.
prisms

Exposure to [28] _____ radiation can result in skin cancer, so sunbed users need to
ultraviolet

take care.

The shortest waves, X-rays and [29] _____ _____, are very hazardous to
gamma   *rays*

humans. Food and medical instruments can be sterilised using [30] _____
gamma

_____ which are also used as tracers in medicine and for treating cancer. X-rays are useful
rays

for examining broken bones and for detecting flaws in other objects.

Science: On Course for GCSE NEAB Edition Teacher's Book © Stanley Thornes (Publishers) Limited 1998

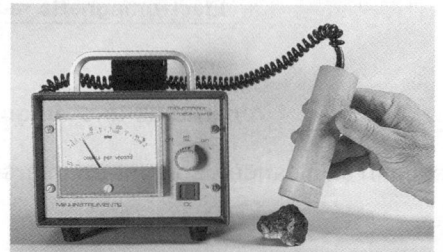

Radioactivity

Do part of this topic for **Single Science** and all of it for **Double Science**.
D = for Double Science only   **H** = for Higher Tier only

📝 Revision notes

Types of radiation

alpha	background radiation	beta	electron
gamma	nucleus	radiation	
electromagnetic radiation	*electron*	*proton*	

There are three main types of ionising radiation emitted when an unstable [1]_____
n u c l e u s

changes to a more stable form. These are called [2]_____ (α), [3]_____ (β)
a l p h a   *b e t a*

and [4]_____ (γ). Complete this table, which compares the properties of these radiations.
g a m m a

ionising radiation	nature	charge	mass	penetration	causes ionisation
alpha particle	two neutrons and two [H5]_____ *protons*	positive	4 × the mass of a proton	absorbed by paper or a few cm of air	intensely
beta particle	fast-moving [H6]_____ *electrons*	negative	1/2000 of the mass of a proton	absorbed by 3 mm of aluminium	weakly
gamma ray	short-wavelength [H7]_____ *electromagnetic* _____ *radiation*	none	none	reduced by several cm of lead	very weakly

We are surrounded by radiation called the [8]_____ _____ ,
b a c k g r o u n d   *r a d i a t i o n*

natural radioactive emissions from the ground, the atmosphere and living things. Radiation ionises

atoms and molecules by removing [9]_____ in collisions. All ionising
e l e c t r o n s

[10]_____ is damaging to living cells and can cause cancer. High doses are used to
r a d i a t i o n

kill cells, including cancer cells and harmful microbes.

Radioactive isotopes have a number of medical and non-medical uses. Alpha and

[11]_____ emitters are used to monitor the thickness of sheet materials. When used as
b e t a

tracers [12]_____ emitters are preferred because their penetration allows them to be
g a m m a

Science: On Course for GCSE NEAB Edition Teacher's Book © Stanley Thornes (Publishers) Limited 1998

detected easily. The fact that they are less ionising than alpha or [13] _____ emitters makes
beta

them safer to use in the body. Radiotherapy also uses [14] _____ emitters to destroy
gamma

cancers.

The atom

alpha	**beta**	**electron**	**isotope**	**mass**
negative	**neutron**	**nucleon**	**nucleus**	**positive**
positively	**proton**	**radioactivity**		
proton	*radionuclide*			

The 'plum pudding' model described an atom as being a solid,

[15] _____ charged sphere with negatively charged
positively

electrons inside it. However, when [16] _____ particles are
alpha

directed at a thin metal film, most of them pass straight through, a small

number being deflected to the side or repelled back. This leads to our

model of an atom as being mainly empty space. Most of the mass is

The arrangement of electrons
in the plum pudding model

in the [17] _____ , a region of intense [18] _____ charge. The negatively
nucleus   *positive*

charged [19] _____ are in orbit around the [20] _____ .
electrons   *nucleus*

The nucleus contains two types of particle, protons and [21] _____ . They have almost
neutrons

identical [22] _____ ; protons have a [23] _____ charge and neutrons have no
mass   *positive*

charge.

The electrons that orbit the [24] _____ each have the same amount of charge as
nucleus

a [25] _____ but of the opposite sign.
proton

In a neutral atom, the positive charge on the nucleus is balanced by the [26] _____ charge
negative

of the orbiting [27] _____ .
electrons

Emission of alpha or [28] _____ radiation results in the formation of a new element as the
beta

numbers of protons and [29] _____ in the nucleus are changed.
neutrons

The total number of neutrons and protons ([30] _____) is the mass number. The most
nucleons

common form of carbon, carbon-12, has six protons and six [31] _____ in the nucleus.
neutrons

Carbon has several [32] _____ , forms of the atom that have the same number of
isotopes

[33] _____ but different numbers of [34] _____ .
protons   *neutrons*

Science: On Course for GCSE NEAB Edition Teacher's Book © Stanley Thornes (Publishers) Limited 1998

Carbon-14 is a radioactive isotope of carbon, known as a radioisotope or

[H35] _____ ; it decays by emitting beta radiation and changing to an isotope of
radionuclide

nitrogen. The emission of beta radiation results in a neutron changing to a [H36] _____ . All
proton

living things have a constant level of carbon-14. When they die the level of carbon-14

[37] *increases*/decreases/*stays the same* as it decays. The age of dead biological material can be

estimated from the level of [38] _____ .
radioactivity

Rocks can also be dated using radioactivity; measurements of the relative amounts of a radioactive

material and the product formed when it decays are used to estimate the age of the rock.

Radioactive decay

counts/s	random			
alpha	gamma	half-life	ionisation	penetration

Radioactive decay is a [39] _____ process; the decay of an individual unstable nucleus
random

cannot be predicted. As the number of undecayed nuclei in a sample decreases, so does the rate of

decay. The average time it takes for half the unstable nuclei in a sample to decay depends only on the

substance and is known as the [H40] _____ . After one half-life the activity of a
half-life

radioactive substance, measured in [41] _____ , can be expected to halve. After two
counts/s

half-lives it is a [H42] *half*/quarter/*eighth* of the original activity and so on.

If a human is exposed to external sources of radiation, beta and [H43] _____ are the most
gamma

dangerous since they have the greatest [H44] _____ . If the source enters the body
penetration

then [H45] _____ is the most dangerous because it causes the most
alpha

[H46] _____ .
ionisation

Nuclear reactors

fission	neutron	radioactive

In a nuclear reactor energy is released when large atomic nuclei are split up in a process called

[DH47] _____ . The nucleus of the atom absorbs a [DH48] _____ and becomes
fission   *neutron*

unstable. When it breaks up it forms two smaller [DH49] _____ nuclei and releases
radioactive

[DH50] _____ that can then cause further fissions.
neutrons

Answers to:
- **Revision Notes**
- **Summary Questions**
- **Formulae, Equations and Calculations**
- **Multiple Choice Questions**

Life processes

1. Nutrition; **2.** Respiration; **3.** Excretion; **4.** Reproduction; **5.** Growth; **6.** Sensitivity;

7. Movement; **8.** Cell membrane; **9.** Cytoplasm; **10.** Nucleus; **11.** Protein coat;

12. Genes; **13.** Cell wall; **14.** Cytoplasm; **15.** Genes; **16.** Nucleus; **17.** Cytoplasm;

18. Cell membrane; **19.** Cell wall; **20.** Tissue; **21.** Contract; **22.** Glandular; **23.** Organ;

24. Organ system; **25.** Surface area; **26.** Alveoli; **27.** Villi; **28.** Diffusion;

29. Cell membrane; **30.** Concentration.

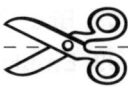

31.

activity	life process
catching rabbits	**nutrition**
flying	**movement**
laying eggs	**reproduction**
breathing out	**excretion**
releasing energy from food	**respiration**
blinking in bright sunlight	**sensitivity**

[6]

1 mark is awarded for each correct answer.
Do not confuse breathing with respiration.

32. (a) Cell membrane: anywhere on the outside
Cytoplasm: anywhere on the inside apart
from the nucleus

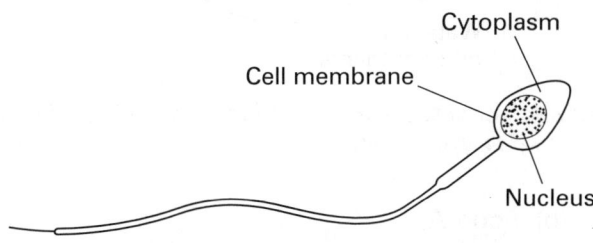

1 mark for each correctly positioned label.   [3]

(b) (i) It controls the cell's activities OR it
contains genetic material.   [1]
(ii) It is where most chemical reactions
occur.   [1]
(iii) It controls the entry and exit of
materials.   [1]
(c) It has a long tail to push against the
water.   [1]
It is streamlined.   [1]

33. *This question is about the way in which a tadpole*
increases the surface area available for absorbing
oxygen.
(a) It is branched.   [1]
The branches increase the surface area   [1]
available for absorption of oxygen.   [1]
(b) It is smaller (OR less active).   [1]
Therefore it does not need as much oxygen
(OR the surface area/volume ratio is higher). [1]

34. *This question is about diffusion, so you should give*
the following ideas:
There would be a net movement of molecules
[1]
from the region of high concentration   [1]
to the region of low concentration.   [1]

| Nutrition

1. Carbohydrate; **2.** Energy; **3.** Protein; **4.** Growth; **5.** Fat; **6.** Energy;

7. Cell membranes; **8.** Gullet; **9.** Liver; **10.** Stomach; **11.** Pancreas; **12.** Small intestine;

13. Large intestine; **14.** Anus; **15.** Gall bladder; **16.** Insoluble; **17.** Soluble;

18. Bloodstream; **19.** Enzymes; **20.** Muscular; **21.** Glandular; **22.** Sugars;

23. Carbohydrase; **24.** Amino acids; **25.** Protease; **26.** Fatty acids; **27.** Lipase;

28. Salivary gland; **29.** Stomach; **30.** Pancreas; **31.** Hydrochloric acid; **32.** Bacteria;

33. Bile; **34.** Gall bladder; **35.** Small intestine; **36.** Acidic; **37.** Alkaline; **38.** Neutralise;

39. Emulsifies; **40.** Surface area; **41.** Lipase; **42.** Sugars; **43.** Amino acids; **44.** Glycerol;

45. Villi; **46.** Surface area; **47.** Large intestine; **48.** Water; **49.** Faeces; **50.** Anus.

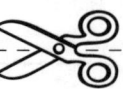

51. (a)

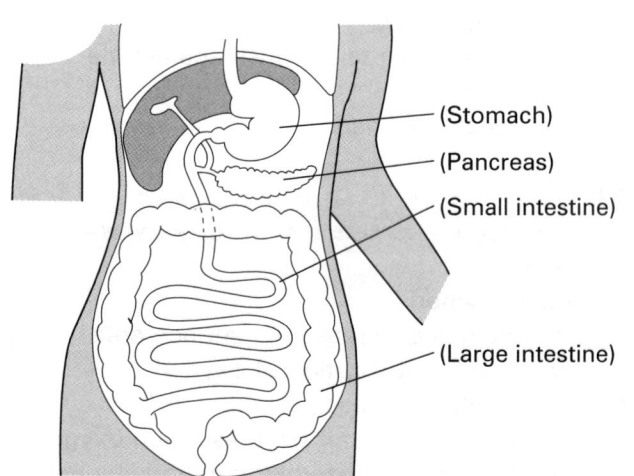

(Stomach)

(Pancreas)

(Small intestine)

(Large intestine)

  (i)   Labelling lines to stomach, small intestine and pancreas   [3]
  (ii)   Labelling lines to pancreas and small intestine   [2]
  (iii) Labelling line to small intestine   [1]
  (iv) Labelling line to large intestine   [1]

1 mark for each correct label.

  (b) It kills the bacteria in food,   [1]
       and provides optimum conditions in which the stomach enzymes work.   [1]
  (c) Muscles in the wall of the digestive system   [1]
      contract to push the food along.   [1]

52. (a)

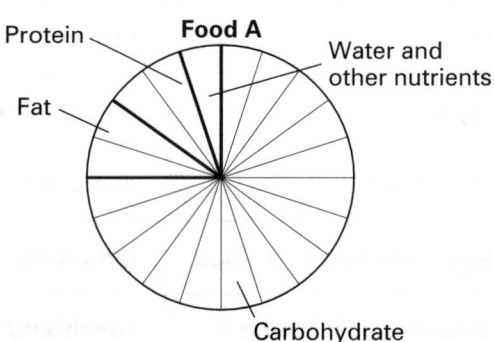

Protein — **Food A** — Water and other nutrients
Fat

Carbohydrate

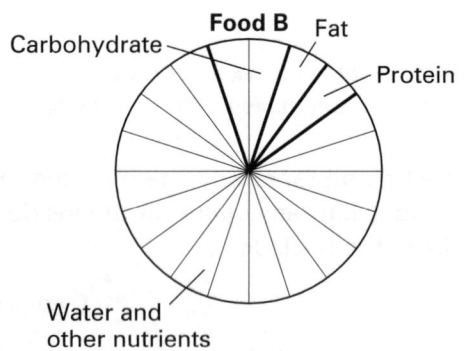

Carbohydrate — **Food B** — Fat — Protein

Water and other nutrients

For each chart, award 1 mark for accurate plotting and 1 mark for labelling.   [4]

  (b) Food A,
      because it contains most protein,   [1]
      which is essential for growth.   [1]

Science: On Course for GCSE NEAB Edition © Stanley Thornes (Publishers) Limited 1988

53. *The idea being tested here is that increasing the surface area increases the rate of absorption.*
The lining of the intestine is folded. [1]
Each fold has projections (called villi). [1]
Together these features increase the surface area of the lining, [1]
leading to a greater rate of absorption of food into the blood. [1]

54. The liver produces bile which emulsifies fats *(breaks down large droplets into smaller ones),* [1]
increasing the surface area for enzymes to act upon. [1]
The pancreas produces the enzyme lipase, [1]
which breaks down the emulsified fats into fatty acids and glycerol. [1]

Science: On Course for GCSE NEAB Edition Teacher's Book © Stanley Thornes (Publishers) Limited 1998

Breathing and respiration

1. Trachea; **2.** Rib; **3.** Bronchus; **4.** Lung; **5.** Diaphragm; **6.** Alveoli; **7.** Rib muscle;
8. Bronchiole; **9.** Muscles; **10.** Downwards; **11.** Volume; **12.** Pressure; **13.** Oxygen;
14. Carbon dioxide; **15.** Energy; **16.** Aerobic; **17.** Anaerobic; **18.** Carbon dioxide;
19. Lactic acid; **20.** Large; **21.** Muscles; **22.** Active transport; **23.** More; **24.** Fatigue;
25. Oxidised; **26.** Oxygen debt; **27.** Lactic acid; **28.** Water; **29.** Surface area;
30. Capillaries.

31. (a) trachea; bronchi; bronchioles; alveoli [4]

1 mark for each.

    (b) (i) 6000 cm^3 ÷ by 10
          = 600 cm^3 [1]
      (ii) (1020 cm^3 ÷ 6000 cm^3) × 100
          = 17% [1]

    (c) (i) The volume of air breathed out
         would increase [1]
         because during exercise the body
         needs more oxygen [1]
         to release energy during respiration. [1]
      (ii) The percentage of carbon dioxide
         would increase [1]
         because the body would be
         respiring faster, [1]
         producing more carbon dioxide. [1]

    (d) *Efficient gaseous exchange needs a large*
      surface area and a good blood supply.
      The alveoli are folded to increase their
      surface area. [2]
      They are surrounded by many blood
      capillaries. [2]

You would also receive credit for stating that
alveoli have thin walls or they are moist.

32. (a) *It takes 4 seconds for one complete breathing*
      cycle.
      Breathing rate = 60 seconds ÷ 4 [1]
               = 15 breaths per minute [1]
    (b) (i) *You have to explain what causes the*
         reduction in pressure in the alveoli.

The intercostal muscles contract,
causing the ribs to rise. [1]
The diaphragm muscles contract,
causing the diaphragm to flatten. [1]
These changes cause the volume of the
thorax to rise, resulting in a decrease in
pressure. [1]

      (ii) *You have to explain what causes the*
         increase in pressure in the alveoli.
         The intercostal muscles relax causing
         the ribs to fall. [1]
         The diaphragm muscles relax causing
         the diaphragm to become
         dome-shaped. [1]
         These changes cause the volume of the
         thorax to fall, resulting in an increase in
         pressure. [1]

33. *The basic idea here is that lactic acid is produced in*
the muscles only when they perform anaerobic
respiration during times of oxygen shortage.
Lactic acid is produced in muscle during
anaerobic respiration, [1]
when the muscles cannot obtain all their
energy needs by aerobic respiration. [1]
The periods of rest in the intermittent
exercise allowed the oxygen levels in the
muscles to be replenished, [1]
resulting in a higher proportion of aerobic
respiration *(which does not produce lactic acid).* [1]

Science: On Course for GCSE NEAB Edition © Stanley Thornes (Publishers) Limited 1988

Circulation and defence

1. Plasma; **2.** Red cells; **3.** Haemoglobin; **4.** Nucleus; **5.** Oxygen; **6.** Lungs; **7.** Organs;

8. White cells; **9.** Platelets; **10.** Organs; **11.** Lungs; **12.** Small intestine; **13.** Urea;

14. Hormones; **15.** Bacteria; **16.** Viruses; **17.** Infection; **18.** Unhygienic; **19.** Toxins;

20. Ingest; **21.** Antitoxins; **22.** Antibodies; **23.** Immune; **24.** Clot; **25.** Hydrochloric acid;

26. Mucus; **27.** Vein; **28.** Atrium; **29.** Valve; **30.** Ventricle; **31.** Artery; **32.** Muscle;

33. Contracts; **34.** Backflow; **35.** Atria; **36.** Ventricles; **37.** Artery; **38.** Vein; **39.** Artery;

40. Artery; **41.** Vein; **42.** Capillary; **43.** Muscle; **44.** Elastic; **45.** High; **46.** Valves;

47. Tissue fluid; **48.** Oxygen; **49.** Carbon dioxide; **50.** Two circulations.

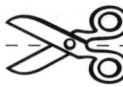

---

51. (a) A: Artery; [1]
because it has the thickest wall OR
because blood is flowing out of it. [1]
B: Capillary; [1]
because it is the narrowest OR
because it connects arteries and veins. [1]
C: Vein; [1]
because blood is flowing into it OR
because it has thinner walls than an
artery. [1]
(b) (i) Tissue fluid [1]
(ii) It passes from the blood through the
capillary walls. [1]
(c) Oxygen [1]
Any food material, e.g. glucose/amino
acid [1]
(d) Carbon dioxide; urea [2]

1 mark for each.

52. (a) There is one circulation between the heart
and the body organs, [1]
and a separate circulation between the
heart and lungs. [1]
(b) Blood is at a low pressure once it has
passed through the lung capillaries. [1]
Returning the blood to the heart allows
its pressure to be raised again to increase
the flow to the organs. [1]

53. *You do not receive marks for a statement without a
reason. For example 'It has an outer layer of tough
fibres' receives no marks.*
(a) Artery
It has an elastic wall so that it can expand
when the heart beats. [1]
The tough outer layer enables it to withstand
the pressure of the heartbeat. [1]
The muscle allows it to constrict to control
the blood supply to organs. [1]
(b) Vein
It has valves to prevent backflow of
blood. [1]
The wall is thinner because blood
pressure is lower than in the artery. [1]
(c) Capillary
Its thin walls allow materials to pass in
and out more easily, [1]
and allow the formation of tissue fluid. [1]
Their small size gives them a large surface
area for the exchange of materials. [1]

Science: On Course for GCSE NEAB Edition Teacher's Book © Stanley Thornes (Publishers) Limited 1998

| Control and co-ordination

1. Eye; 2. Ear; 3. Skin; 4. Nose; 5. Sclera; 6. Ciliary muscle; 7. Suspensory ligament;

8. Cornea; 9. Pupil; 10. Lens; 11. Iris; 12. Optic nerve; 13. Retina;

14. Suspensory ligament; 15. Sclera; 16. Iris; 17. Retina; 18. Cornea; 19. Lens;

20. Optic nerve; 21. Addiction; 22. Withdrawal symptoms; 23. Brain; 24. Cancer;

25. Emphysema; 26. Depressant; 27. Liver; 28. Impulses; 29. Sensory neurons;

30. Relay neurons; 31. Synapse; 32. Chemicals; 33. Motor neurons; 34. Effectors;

35. Glands; 36. Response; 37. Reflex action; 38. Stimulus; 39. Receptor; 40. Co-ordinator;

41. Effector; 42. Response; 43. Motor neuron; 44. Sensory neuron; 45. Receptor;

46. Cornea; 47. Ciliary muscles; 48. Suspensory ligaments; 49. Lens; 50. Retina.

51. (a) A: Sclera   [1]
       B: Pupil   [1]
       C: Iris (cornea would be correct)   [1]
    (b) (i) The pupil would become smaller OR
             the iris would become bigger.   [1]
        (ii) Receptors in the eye detect an
             increase in light intensity.   [1]
             Impulses pass along a sensory neurone
             (OR the optic nerve)   [1]
             to the brain.   [1]
             The brain co-ordinates the response.   [1]

52. (a) Spirits (40% ABV)   [1]
    (b) 4 pints of beer is 8 half pints which contain
        8 units, 2 double measures contain
        4 units,   [1]
        Answer: 12 units   [1]
    (c) Alcohol slows down a person's
        reactions.   [1]
        Therefore the driver would respond more
        slowly to an emergency.   [1]

53. (a) *You have to explain how Parts A and B change*
        the shape of C.
        Part A (ciliary muscles) relaxes,   [1]
        The tension in Part B (suspensory
        ligaments) is increased,   [1]

        owing to elastic recoil of the wall of the eyeball.

        so the lens is now under tension and is
        pulled into a thinner shape,   [1]
        which gives the longer focal length needed to
        focus on distant objects.   [1]

    (b) Information is transmitted along neurons
        by electrical impulses   [1]
        and across synapses   [1]
        by chemical transmitter substances.   [1]

Homeostasis

1. Sweat; 2. Urine; 3. Carbon dioxide; 4. Urea; 5. Lungs; 6. Skin; 7. Carbon dioxide;
8. Urea; 9. Liver; 10. Bladder; 11. Skin; 12. Hormones; 13. Glands; 14. Plasma;
15. Pancreas; 16. Glucagon; 17. Diabetes; 18. Glucagon; 19. Glycogen; 20. Filtration;
21. Sugar; 22. Water; 23. Urea; 24. ADH; 25. Increases;
26. Thermoregulatory centre; 27. Dilate; 28. Constrict; 29. Evaporates; 30. Respiration.

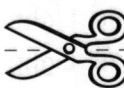

31. (a) *Many students confuse the jobs of the liver,*
    kidney and bladder.
    (i)   E *(skin)*   [1]
    (ii) A *(lungs)*   [1]
    (iii) D *(liver)*   [1]
    (iv) B *(kidney)*   [1]
    (v)   C *(bladder)*   [1]
    (b) *Remember that on hot days we sweat more*
        to cool the body.
        An increase in outside temperature may
        lead to an increase in body temperature. [1]
        The body responds by increasing the rate
        of sweating *(to cool the body).*   [1]
        We drink more to replace the water lost
        by the increased sweating.   [1]

32. (a) (i) Pancreas   [1]
        (ii) High blood sugar levels   [1]
    (b) *You need to explain what causes the increase in*
        blood sugar and how insulin reduces blood sugar.
        Starch in the meal is digested to form
        sugars,   [1]
        which pass into the blood causing an
        increase in blood sugar levels.   [1]
        This stimulates the pancreas to produce
        insulin.   [1]
        Insulin stimulates liver cells   [1]
        to remove sugar from the blood and convert
        it into glycogen.   [1]

Photosynthesis and growth

1. Support; 2. Anchorage; 3. Water; 4. Photosynthesis; 5. Carbon dioxide; 6. Water;
7. Light; 8. Chlorophyll; 9. Chloroplasts; 10. Glucose; 11. Oxygen; 12. Respiration;
13. Chloroplasts; 14. Cytoplasm; 15. Cell membrane; 16. Nucleus; 17. Vacuole;
18. Cell sap; 19. Cell wall; 20. Energy; 21. Growth; 22. Cellulose; 23. Starch;
24. Nitrates; 25. Protein; 26. Carbon dioxide; 27. Energy; 28. Chlorophyll; 29. Oxygen;
30. Light; 31. Carbon dioxide; 32. Temperature; 33. Gravity; 34. Tip; 35. Hormones;
36. Light; 37. Water; 38. Unequal; 39. Hormones; 40. Weeds; 41. Cutting; 42. Roots;
43. Fruits; 44. Nitrate; 45. Protein; 46. Photosynthesis; 47. Purple; 48. Enzymes;
49. Yellow; 50. Limiting factor.

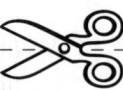

51. (a) A, C and D *(the parts light can reach)* [1]
    (b) Only the parts that received light [1]
        could photosynthesise to produce
        glucose. [1]
    (c) carbon dioxide + water $\xrightarrow[\text{chlorophyll}]{\text{light}}$ glucose
        + oxygen [3]

Award 1 mark for carbon dioxide + water; 1 mark for glucose + oxygen; 1 mark for light and chlorophyll.

52. *This question is about phototropism in plant stems.*
    (a) The shoots grew [1]
        mainly at the tips [1]
        and towards the light. [1]

(b) Growth is controlled by hormones. [1]
    Light on one side of the stem results in a
    greater concentration of hormone on the
    other side. [1]
    The greater concentration of hormone on
    the 'dark' side causes the cells on that side
    to grow faster. [1]

53. Plant A:
    The deficient mineral ion is nitrate, [1]
    which is needed for protein synthesis. [1]
    Plant B:
    The deficient mineral ion is phosphate, [1]
    which is needed for reactions in photosynthesis
    and respiration. [1]

Water relations

1. Root hair; **2.** Xylem; **3.** Transpiration; **4.** Evaporates; **5.** Stomata; **6.** Transpiration;
7. Hot; **8.** Windy; **9.** Low; **10.** Wax; **11.** Thicker; **12.** Guard cells; **13.** Carbon dioxide;
14. Lower; **15.** Less; **16.** Cooler; **17.** Wilting; **18.** Phloem; **19.** Partially permeable;
20. Concentration gradient; **21.** Osmosis; **22.** Starch; **23.** Insoluble; **24.** Active uptake;
25. Energy; **26.** Root hairs; **27.** Air spaces; **28.** Osmosis; **29.** Cell wall; **30.** Turgor.

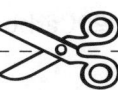

31. (a) Water evaporates from the leaves,   [1]
    causing water to be pulled   [1]
    up the xylem vessels which open into the
    water in the beaker.   [1]
    (b) $(4.5 + 5.2 + 4.8 + 4.7 + 5.0)$ cm $\div 5$ [1]
    $= 4.8$ cm   [1]
    (c) The bubble would move faster   [1]
    because the water would evaporate faster
    from the leaves in moving air.   [1]
    (d) *The clear plastic bag would trap water vapour*
    and therefore increase the humidity around the
    leaves.
    The bubble would move more slowly   [1]
    because the water would evaporate more
    slowly from the leaves in humid air.   [1]

32. (a) The plants had wilted   [1]
    because they had lost water   [1]
    faster than they could absorb it from the
    dry soil.   [1]

    (b) It is colder at night,   [1]
    and stomata close in the dark *(reducing the*
    escape of water vapour),   [1]
    so the rate of evaporation from the plants
    would have been slower.   [1]

33. *This question is about osmosis, which is a special*
    case of diffusion, so you should give the following
    ideas:
    The cells of the sultana have partially
    permeable membranes   [1]
    that allow water molecules to pass through
    quickly, but sugar molecules only slowly.   [1]
    Because there is a high concentration of
    sugars in the sultanas   [1]
    there is a low concentration of water
    molecules,   [1]
    so there is a net diffusion of water
    molecules into the sultana.   [1]

    You would also gain credit if you stated that water
    diffuses from a weak solution to a concentrated
    solution.

Science: On Course for GCSE NEAB Edition Teacher's Book © Stanley Thornes (Publishers) Limited 1998

The environment

1. Water; **2.** Nutrients; **3.** Light; **4.** Oxygen; **5.** Carbon dioxide; **6.** Temperature;

7. Nutrients; **8.** Breeding; **9.** Predators; **10.** Prey; **11.** Rise; **12.** Fall; **13.** Non-renewable;

14. Fossil fuels; **15.** Combustion; **16.** Pollution; **17.** Carbon dioxide;

18. Sulphur dioxide; **19.** Acidic; **20.** Leaves; **21.** Acidic; **22.** Fertilisers; **23.** Pesticides;

24. Competition; **25.** Microbes; **26.** Respiration; **27.** Oxygen; **28.** Methane;

29. Carbon dioxide; **30.** Radiation.

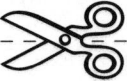

31. (a) light; nutrients [2]

1 mark for each.

    (b) food; space for breeding [2]

1 mark for each.

    (c) A rise in the population of small animals *(prey)* would provide more food for the fish *(predators)*. [1]
The number of predators would increase, causing a decrease in the number of prey, [1]
which in turn would result in fewer predators. [1]

32. (a) It is caused by the combustion [1]
of fossil fuels, [1]
which releases sulphur dioxide. [1]
This dissolves in rain to form an acid. [1]
    (b) (i) 40% [1]
        (ii) 30% [1]
    (c) Oak trees lose their leaves every winter whereas pine trees keep theirs. [1]
Pine leaves are therefore exposed to acid for longer, and therefore more likely to be damaged. [1]

33. *The crucial point in eutrophication is that it is the respiration of decay microbes that causes reduction in the oxygen content of the water.*
Carbohydrate is a food for microbes. [1]
Therefore the population of microbes will increase. [1]
Their respiration [1]
will deplete the oxygen content of the water, [1]
and larger animals may suffocate because of this oxygen reduction. [1]

34. (a) *There are two effects of this method of clearing land: deforestation and combustion.*
Deforestation increases the carbon dioxide content of the air [1]
because the trees are no longer taking in carbon dioxide for photosynthesis. [1]
Burning the trees also raises the carbon dioxide content of the atmosphere, [1]
since carbon dioxide is a by-product of combustion. [1]
    (b) *Many developing countries cannot afford fertilisers.*
Cropping removes mineral ions from the soil. [1]
If these are not replaced the soil is eventually unsuitable for plant growth. [1]
One 'cheap' solution is the use of crop rotation, [1]
which replaces mineral ions naturally. [1]

Science: On Course for GCSE SEG Edition © Stanley Thornes (Publishers) Limited 1988

Energy flow and nutrient cycles

1. Producers; **2.** Photosynthesis; **3.** Consumers; **4.** Energy; **5.** Radiation; **6.** A; **7.** A; **8.** B;

9. A; **10.** C; **11.** Detritus feeders; **12.** Digestion; **13.** Microbes; **14.** Moist; **15.** Warm;

16. Oxygen; **17.** Sewage works; **18.** Compost heaps; **19.** Photosynthesis;

20. Carbohydrate; **21.** Respiration; **22.** Respiration; **23.** Microbes; **24.** Respiration;

25. Faeces; **26.** Movement; **27.** Heat; **28.** Ammonium compounds;

29. Nitrifying bacteria; **30.** Nitrates.

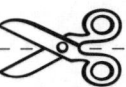

31. (a) *1 mark for each correctly placed organism.*   [3]

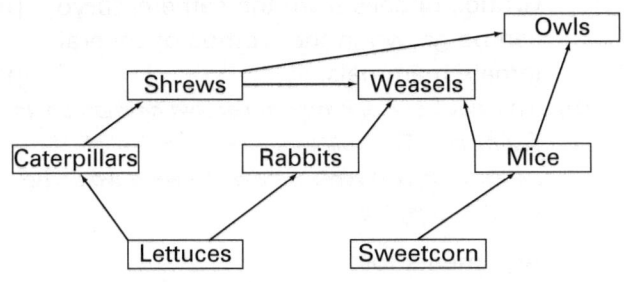

(b) (i) EITHER lettuce OR sweetcorn   [1]
   (ii) EITHER caterpillars OR rabbits OR mice   [1]
   (iii) EITHER shrews OR weasels OR owls [1]

(c) (i) *Several caterpillars feed on each lettuce, so the bar for the first trophic level is narrower than that for the second.*

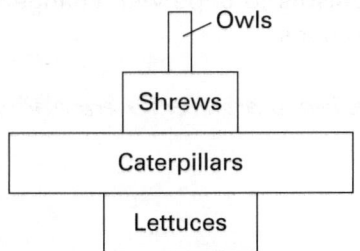

1 mark for the correct shape and 1 mark for all labels being present and in the correct order.   [2]

(ii)

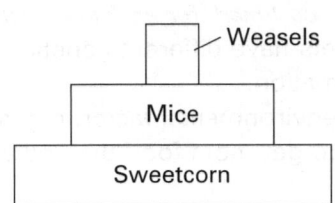

1 mark for the correct shape and 1 mark for all labels being present and in the correct order.   [2]

(d) (i) Microbes   [1]
       digest dead material.   [1]
   (ii) Presence of water,   [1]
       warmth,   [1]
       and oxygen.   [1]

32. *Remember that all living organisms respire.*
   A: Photosynthesis   [1]
   B: Decay   [1]
   C: Respiration   [1]
   D: Respiration   [1]

Science: On Course for GCSE NEAB Edition Teacher's Book © Stanley Thornes (Publishers) Limited 1998

Variation and selection

1. Genes; **2.** Environmental; **3.** Genetic; **4.** Asexual; **5.** Identical; **6.** Fertilisation;
7. Sexual; **8.** Different; **9.** Clones; **10.** Chromosomes; **11.** Alleles; **12.** Mutations;
13. Radiation; **14.** Cancer; **15.** Cuttings; **16.** Identical; **17.** Artificial selection;
18. Characteristics; **19.** Alleles; **20.** Mitosis; **21.** Meiosis; **22.** Halves; **23.** Meiosis;
24. Parent; **25.** Alleles; **26.** Tissue culture; **27.** Embryo; **28.** DNA; **29.** Proteins;
30. Bacteria.

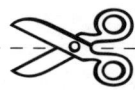

---

31. (a) *Award 2 marks if all plots are correct.*
Subtract 1 mark for each incorrect plot. [2]

(b) Limpets have different genetic
information. [1]
One environmental factor, e.g. some were
able to get more food than others. [1]

32. (a) 3709 − 2547 = 1162 litres [1]

(b) *This question requires an answer in terms of
artificial selection.*
Farmers select the highest-yielding cows [1]
to breed the next generation, [1]
and they reselect for yield every year. [1]

33. (a) *EITHER tissue culture:*
Many small groups of cells from chosen
plants [1]
can be grown in special media to form
young plants; [1]

OR embryo transplants:
Groups of cells from the same embryo [1]
can be grown in the wombs of several
female mammals. [1]

(b) *You must use arguments for and against to gain
full marks. The following are examples of
acceptable answers. Similar answers would be
equally acceptable.*

Arguments for:
All the organisms will have the desired
characteristics, [1]
and are therefore more valuable. [1]

Arguments against:
Selective breeding reduces the number of
alleles, [1]
which might restrict attempts to breed
organisms to cope with changed
conditions. [1]

Moral and religious arguments are also acceptable.

Science: On Course for GCSE NEAB Edition © Stanley Thornes (Publishers) Limited 1988

Inheritance and evolution

1. Nucleus; 2. Chromosomes; 3. Alleles; 4. In pairs; 5. Single; 6. XX; 7. XY; 8. X; 9. Y;
10. XX; 11. XY; 12. Girl–boy; 13. Dominant; 14. Recessive; 15. Homozygous;
16. Heterozygous; 17. Carriers; 18. h; 19. h; 20. Hh; 21. hh; 22. Hh; 23. hh; 24. 1:1;
25. c; 26. C; 27. CC; 28. Cc; 29. CC; 30. Cc; 31. none; 32. Womb; 33. Hormones;
34. Pituitary; 35. Fertility drug; 36. Contraceptive drugs; 37. FSH; 38. Oestrogens;
39. FSH; 40. LH; 41. Pituitary; 42. LH; 43. FSH; 44. Oestrogen; 45. FSH; 46. Fossils;
47. Decay; 48. Evolution; 49. Adaptations; 50. Variations.

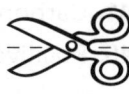

51. (a) genes (OR alleles) [1]
    (b) XY [1]
    (c) XX [1]

52. (a) There are fossil remains [1]
       of parts that did not decay/hard parts. [1]
       We know the age of the rocks the fossils are
       found in. [1]
    (b) They are larger. [1]
       Their legs are longer in proportion to their
       bodies. [1]
    (c) e.g. new, faster predators evolved. [1]
    (d) *You need to explain this in terms of mutation,
       variation and natural selection.*
       Some animals had genetic information for
       developing longer legs, [1]
       which possibly arose via mutation. [1]
       Animals with longer legs could run
       faster [1]
       and escape more easily from predators, [1]
       so they survived to breed. [1]
       Their genes were passed to the next
       generation whereas those of smaller, slower
       animals were not. [1]

53. *In this type of answer you must give **both**
    advantages and disadvantages, not simply
    concentrate on one or the other. You should then
    give your reasoned opinion.*
    Advantages include:
    They can prevent unwanted pregnancy.
    They can treat infertility in females.
    Disadvantages include:
    They may promote promiscuity.
    There is an increased risk of spreading sexual
    diseases.
    There is a risk of multiple pregnancy.
    Health risks, e.g. increased chance of blood
    clots.
    Opinion: e.g.
    Benefits are highly positive and outweigh
    disadvantages which carry a small risk. [6]

    1 mark for each valid point up to a maximum of 6.

Science: On Course for GCSE NEAB Edition Teacher's Book © Stanley Thornes (Publishers) Limited 1998

| Metals

1. Metal; **2.** Boiling points; **3.** Soft; **4.** Carbon; **5.** Reactivity; **6.** Top; **7.** Paraffin oil;

8. Oxygen; **9.** Oxide; **10.** Hydroxide; **11.** Hydrogen; **12.** Salt; **13.** Hydrogen; **14.** Copper;

15. Displacement; **16.** Copper; **17.** Iron; **18.** Zinc; **19.** Magnesium; **20.** Rock;

21. Electrolysis; **22.** Reduction; **23.** Uncombined; **24.** Blast; **25.** Coke; **26.** Air;

27. Carbon; **28.** Carbon dioxide; **29.** Reduced; **30.** Reducing agent;

31. Acidic impurities; **32.** Slag; **33.** Reducing agent; **34.** Aluminium oxide;

35. Electrolysis; **36.** Cryolite; **37.** Carbon; **38.** Aluminium; **39.** Oxygen; **40.** Burn;

41. Carbon dioxide; **42.** Electrolysis; **43.** Electrolyte; **44.** Negative electrode;

45. Positive electrode; **46.** Ions; **47.** Gain; **48.** Reduction; **49.** Lose; **50.** Oxidation.

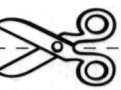

51. (a) Zinc is above silver in the reactivity
        series.   [1]
    (b) $Zn(s) + \dots 2. AgNO_3(aq)$
         $\rightarrow \dots Zn(NO_3)_2(aq) \dots + \dots 2\ Ag(s) \dots$ [1]
    (c) Add nickel to solutions of the nitrates of the
         different metals in separate tubes.   [1]
         Metals below nickel in the reactivity series
         will be displaced.   [1]
         Metals above nickel will not be displaced /
         no reaction.   [1]

52. (a) (i) Hydrogen   [1]
        (ii) Put a lighted splint into a test tube of
             gas.   [1]
             Gas burns with a squeaky pop.   [1]
    (b) **Y Z W X**   [3]

*Award 1 mark if **Y** is before **Z** in the list; 1 mark if **Z**
is before **W**; 1 mark if **W** is before **X**.*

53. (a) **D**   [1]
    (b) **A**   [1]
    (c) Reaction: **B**   Substance reduced: **CO$_2$**
         Reaction: **C**   Substance reduced: **Fe$_2$O$_3$** [4]

1 mark for each answer.

    (d) Calcium oxide reacts with silicon(IV) oxide
         to produce calcium silicate (slag).   [1]
         Slag can be tapped off from the furnace.   [1]
         Enables the furnace to act continuously.   [1]

54. (a) Carbon electrodes burn away.   [1]
    (b) Negative
         electrode $\Big\}$ $Al^{3+} + \dots 3e^- \dots \rightarrow .Al \dots$ [2]

         Positive
         electrode $\Big\}$ $\dots 2\ O^{2-} \rightarrow \dots O_2 + \dots 4e^- \dots$ [2]

*In each equation there is 1 mark for the correct
species and 1 mark for balancing.*

    (c) Oxidation: loss of electrons. Reduction:
         gain of electrons.   [1]
         Al^{3+} gains electrons and so is reduced. [1]
         O^{2-} loses electrons and so is oxidised. [1]

Science: On Course for GCSE NEAB Edition © Stanley Thornes (Publishers) Limited 1988

Acids, bases and salts

1. Acids; **2.** Sour; **3.** Hydrogen; **4.** Salt; **5.** Acids; **6.** Alkalis; **7.** Neutral; **8.** Universal;
9. Meter; **10.** Weak; **11.** Strong; **12.** Hydrogen; **13.** Carbon dioxide; **14.** Limewater;
15. Sulphate; **16.** Salt; **17.** Neutralisation; **18.** Water; **19.** Precipitation; **20.** Chloride;
21. Nitric; **22.** Sodium; **23.** Sulphuric; **24.** Sulphate; **25.** Water; **26.** Ammonia;
27. Carbon dioxide; **28.** Sulphuric; **29.** Nitric; **30.** Water.

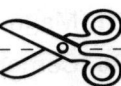

- -

31. (a) Ammonium sulphate [1]
    (b) Carbon dioxide [1]
    (c) Barium sulphate [1]
    (d) Barium chloride [1]

32. (a) A base is a metal oxide. [1]

    (b) (i) $O^{2-}(s) + H_2O(l) \rightarrow 2OH^-(aq)$ [2]

1 mark for correct formulae and 1 mark for balancing.

        (ii) $H^+(aq) + OH^-(aq) \rightarrow H_2O(l)$ [2]

1 mark for correct left-hand side and 1 mark for correct right-hand side.

Science: On Course for GCSE NEAB Edition Teacher's Book © Stanley Thornes (Publishers) Limited 1998

Rocks in the Earth

1. Minerals; **2.** Diamond; **3.** Magma; **4.** Igneous; **5.** Crystals; **6.** Slowly; **7.** Intrusive;

8. Crystals; **9.** Quickly; **10.** Extrusive; **11.** Sedimentary; **12.** Older; **13.** Shale; **14.** Soft;

15. High; **16.** High; **17.** Metamorphic; **18.** Marble; **19.** Calcium carbonate; **20.** Slate;

21. Intrusion; **22.** Igneous; **23.** Sedimentary; **24.** Metamorphic; **25.** Extrusive;

26. Igneous; **27.** Intrusive; **28.** Igneous; **29.** Sedimentary; **30.** Metamorphic;

31. Magma; **32.** Weathering; **33.** Erosion; **34.** Transported; **35.** Deposited;

36. Cementation; **37.** Burial; **38.** Recrystallisation; **39.** Crystallisation; **40.** Melting;

41. Crystallisation; **42.** Erosion; **43.** Transportation; **44.** Deposition; **45.** Burial;

46. Cementation; **47.** Sedimentary; **48.** Igneous; **49.** Fossils; **50.** Radioactivity.

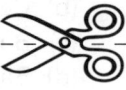

---

51. (a) Crystals, does not fizz with acid, does not split into layers, crystals mixed up. [3]

There are four points in the mark scheme. One point scores 1 mark, two or three points score 2 marks and all four points score 3 marks.

    (b) Metamorphic [1]
        Crystals in layers [1]
    (c) Running water will remove sharp edges from pieces of rock. [1]
    (d) Granite [1]
        Slate [1]

52. (a) Bubbles of colourless gas. [1]
        calcium carbonate + hydrochloric acid
        → calcium chloride + carbon dioxide
        + water [2]
    (b) Making cement [1]

53. (a) Rocks covering the granite have eroded away leaving granite exposed. [1]

Granite is slower to erode than many of the softer sedimentary rocks around it, e.g. limestone, sandstone.

    (b) High temperatures by day cause outer rock to expand. [1]
        At night temperature falls, causing the outer rock to cool and contract. [1]
        Rock is a poor conductor of heat, so the temperature inside it does not change as much as it does near the surface. [1]

Expansion and contraction of the outer layers cause strain inside the rock, resulting in the outer layers cracking off. [1]

    (c) Rainwater penetrates into small cracks in the outer layer of the rock. [1]
        In cold weather water freezes in these cracks. [1]
        When water freezes it expands. [1]
        Expansion in the outer layers causes strain inside the rock, resulting in outer layers cracking off. [1]

Chemical weathering is beyond the scope of GCSE syllabuses but acid in water is involved in the weathering of the rocks on Dartmoor. Obviously any answer in terms of chemical weathering would receive full credit.

54. (a) Rocks are in layers. [1]
        Fossils in rock. [1]

Sometimes there are ripple marks in sedimentary rocks.

    (b) On the diagram, where the limestone and granite are in contact. [1]

The high temperatures and high pressures at this point turn limestone into marble.

    (c) **R–S** goes through more rock layers than **X–Y**. [2]

Chemicals from oil

1. Organic; **2.** Petroleum; **3.** Gas; **4.** Water; **5.** Gas; **6.** Non-porous rocks; **7.** Fossil;

8. Oxygen; **9.** Carbon; **10.** Hydrocarbons; **11.** Fractional distillation; **12.** Boiling points;

13. Crude oil vapour; **14.** Petroleum gases; **15.** Petrol; **16.** Diesel; **17.** Lubricating oil;

18. Bitumen; **19.** Top; **20.** Higher; **21.** Alkanes; **22.** Methane; **23.** Covalent;

24. Saturated; **25.** Increases; **26.** Viscosity; **27.** More difficult; **28.** Carbon dioxide;

29. Cracking; **30.** Catalyst; **31.** Unsaturated; **32.** Alkenes; **33.** Ethene;

34. Polymerisation; **35.** Polymers; **36.** Monomers; **37.** Poly(propene);

38. Poly(vinyl chloride); **39.** Vinyl chloride; **40.** Electricity.

---

41. (a) (i)   150 °C   [1]
          (ii) As the number of carbon atoms
                increases, the boiling point
                increases.   [1]
      (b) C_8H_{18} [1]
      (c) Hexane   [1]
      (d) pentane + air (plentiful supply)
          $\rightarrow$ carbon dioxide + water   [1]

Burning any carbon-based fuel in a limited supply of oxygen produces carbon monoxide which is a very toxic gas.

42. (a) First test tube of gas collected contains
           largely air pushed out of the apparatus.   [1]

The reaction taking place is an example of cracking. Long chains are being broken into short molecules which have much greater economic value.

      (b) ....1.C_2H_4. + ...3...O_2
          $\rightarrow$ ..2..CO_2.. + ...2...H_2O [2]

1 mark is awarded for the correct formula for carbon dioxide and 1 mark for balancing.

      (c) (i) Poly(ethene)   [1]
          (ii)

$$\left[\begin{array}{cc} \overset{\displaystyle H}{|} & \overset{\displaystyle H}{|} \\ -C- & -C- \\ \underset{\displaystyle H}{|} & \underset{\displaystyle H}{|} \end{array}\right]_n$$

                                                                   [2]

1 mark is for showing the double bond between two carbon atoms becoming a single bond. The second mark is for an indication that the molecules are connected in a chain.

          (iii) Children's toys: non-toxic, no sharp
                edges, can be brightly coloured.   [2]
                Gardener's twine: strong, does not
                rot, does not stretch.   [2]

There are a variety of possible answers to these two parts. Three possible answers have been included.

43. (a) Cyclobutane contains only single covalent
           bonds between carbon atoms
           (saturated).   [1]
           Butene contains a double bond between
           two carbon atoms (unsaturated).   [1]
      (b) Butane   [1]

$$\begin{array}{ccccccc}
 & H & & H & & H & & H & \\
 & | & & | & & | & & | & \\
H- & C & - & C & - & C & - & C & -H \\
 & | & & | & & | & & | & \\
 & H & & H & & H & & H &
\end{array}$$

                                                                   [2]

You will lose a mark here if you miss off the hydrogen atoms or do not put bonds between the carbon atoms.

Science: On Course for GCSE NEAB Edition Teacher's Book © Stanley Thornes (Publishers) Limited 1998

| The Earth and its atmosphere

1. Crust; **2.** Mantle; **3.** Convection; **4.** Outer core; **5.** Inner core; **6.** Nickel; **7.** Liquid;
8. More dense; **9.** Forces; **10.** Plates; **11.** Tectonics; **12.** Fossils; **13.** Slide; **14.** Faults;
15. Boundaries; **16.** Mantle; **17.** Magnetic field; **18.** Oceanic; **19.** Continental;
20. Mantle; **21.** Subduction; **22.** Volcanoes; **23.** Mixture; **24.** Oxygen; **25.** Respire;
26. Burned; **27.** Photosynthesis; **28.** Increased; **29.** Greenhouse effect;
30. Global warming.

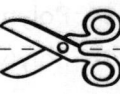

---

31. (a) Carbon dioxide   [1]

The important word in this question is **compound.**

    (b) Sulphur dioxide – from the burning of coal
        and oil   [1]
        Oxides of nitrogen – from internal
        combustion engines in cars etc.   [1]

32. (a) Iceland is on the Mid-Atlantic Ridge.   [1]
        Two plates are moving apart.   [1]
        New igneous rocks formed where rocks
        from the magma come to the surface.   [1]
    (b) Nazca and South American plates
        colliding.   [1]
        Rocks are forced upwards   [1]
        as one plate moves over the other.   [1]

33. *For 6 marks the answer should be detailed and
include the following points:*
Volcanic origin of gases in the atmosphere.
Temperature of the surface cools so liquid water
is formed.
Build-up of oxygen by photosynthesis.
Role of nitrifying and denitrifying bacteria.
Formation and role of ozone layer.
Balance between oxygen and carbon dioxide
established.

*For 4 marks, either a detailed account of four of
the above or incomplete answers to six.*
*For 2 marks, either a detailed account of two of
the above or an incomplete answer to four.*

Science: On Course for GCSE NEAB Edition © Stanley Thornes (Publishers) Limited 1988

Rates of chemical reactions

1. Very fast; 2. Explosion; 3. Very slow; 4. Decreases; 5. Costs; 6. Collision;

7. Effective; 8. Activation energy; 9. Speed up; 10. Halve; 11. Double; 12. Speeds up;

13. Collisions; 14. Cooling; 15. Temperature; 16. Decrease; 17. Increase; 18. Pressure;

19. Collisions; 20. Faster than; 21. Surface area; 22. Explosion; 23. Start; 24. Chlorine;

25. Catalyst; 26. Stays the same; 27. The same mass of product; 28. Surface;

29. Compound; 30. Activation energy; 31. Finished; 32. Reactants; 33. Fastest;

34. Concentrated; 35. The same; 36. Less than; 37. Enzymes; 38. Faster;

39. Denatured; 40. Irreversibly; 41. Amylase; 42. Saliva; 43. Alcohol;

44. Carbon dioxide; 45. Fermentation; 46. Limewater; 47. Carbon dioxide; 48. Yoghurt;

49. Lactose; 50. Lactic acid.

---

51. (a) Sulphur is precipitated during the experiment. At some point the solution becomes just so cloudy that the cross cannot be seen through it. [1]
    (b) Water and sodium thiosulphate are mixed and then hydrochloric acid added. [1]
    (c) The best time to take the temperature is soon after the addition of hydrochloric acid. [1]

At high temperatures, changes of temperature become more significant. Here temperatures can be taken when hydrochloric acid is added and again at the end of the experiment. An average can then be used.

    (d) Timing should be started when hydrochloric acid is added. [1]
    (e) Experiment number 2 [1]
        This experiment was carried out at the highest temperature and temperature has a very big effect. [1]
    (f) (i) 3, 4, 5 and 7 [1]
        (ii) Correct order: 5, 3, 7, 4 (decreasing rate) [1]
            The concentration of sodium thiosulphate decreases. [1]
            There are more collisions between thiosulphate ions and hydrogen ions in concentrated solutions. [1]
    (g) (i) 2, 3, 8, 9 [1]
        (ii) Correct order: 2, 8, 9, 3 (decreasing rate) [1]

    As the temperature increases, the particles move faster, [1]
    resulting in more effective collisions. [1]
    (h) The cross is viewed through a greater depth of solution. [1]

52. (a) Oxygen [1]

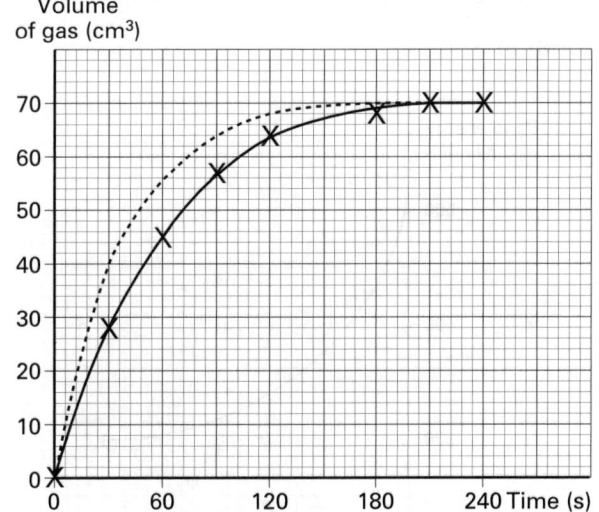

    (b) (i) Plotting points [2]
            Curve drawn [1]
        (ii) 0.5 g [1]

The mass of the catalyst remains unchanged.

        (iii) Sketch graph.
            The graph should be steeper; [1]
            reach the same maximum (70 cm³). [1]

Science: On Course for GCSE NEAB Edition Teacher's Book © Stanley Thornes (Publishers) Limited 1998

| Energy changes in reactions

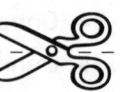

1. Oxidation; **2.** Energy; **3.** Fuel; **4.** Oxygen; **5.** Carbon; **6.** Hydrogen; **7.** Exothermic; **8.** Endothermic; **9.** Bonds; **10.** Bond breaking; **11.** Bond making; **12.** Bond breaking; **13.** Bond making; **14.** Bond making; **15.** Bond breaking; **16.** Activation energy; **17.** Catalyst; **18.** Energy level diagrams; **19.** Energy; **20.** Reactants; **21.** Activation energy; **22.** Products; **23.** Energy change; **24.** Energy; **25.** Reactants; **26.** Activation energy; **27.** Products; **28.** Energy change; **29.** Exothermic; **30.** Products.

---

31. (a) $C=C$ [1]
    (b) $C=C$ [1]
    (c) As the group is descended the bond energy increases from F_2 to Cl_2 and then decreases.   [1]

This pattern is not easy to see because it is not a simple trend.

    (d) 876 kJ/mol   [1]

The additional bond energy from C—C to C=C is 264 kJ/mol. It is sensible to predict that the additional bond energy from C=C to C≡C is 264 kJ/mol.

    (e) (i)

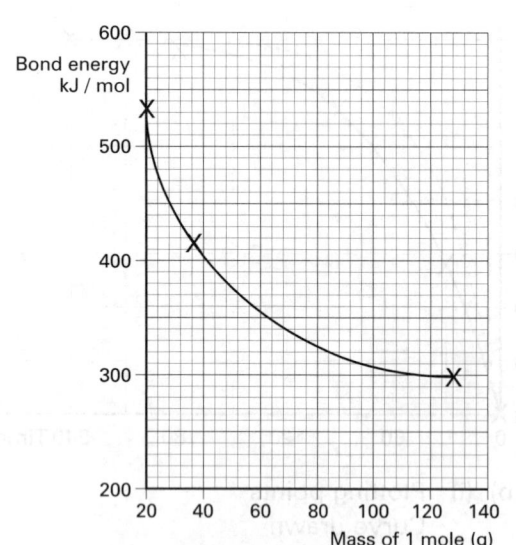

        Correct plotting   [2]

1 mark awarded instead of 2 if one point wrongly plotted.

        Good curve drawn.   [1]
      (ii) In the range 325–45 kJ/mol   [1]

The answer will vary according to the line drawn on the graph.

    (f) (i) $436 + 193$ kJ $= 629$ kJ   [1]
      (ii) If the correct answer is x kJ/mol, energy required to break bonds $= 629$ kJ/mol. The difference between the energy required and the energy given out is the energy released.

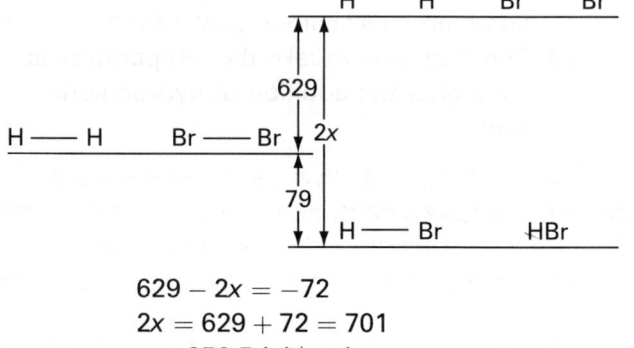

$$629 - 2x = -72$$
$$2x = 629 + 72 = 701$$
$$x = 350.5 \text{ kJ/mol} \quad [2]$$

You should not be surprised that the value obtained is not the same as the value in the data table (on page 88). These are only average values.

32. (a) Energy required to break bonds
        $= 151 + 242 = 393$ kJ   [1]
        Energy released when bonds form
        $= 2 \times 202 = 404$ [1]
        Energy change $= 393 - 404 = -11$ kJ   [1]
    (b) The energy released is greater than the energy required. The reaction is exothermic.   [1]

A negative value in the calculation shows the reaction is exothermic. Usually we use a convention of making energy required positive and energy released negative.

Science: On Course for GCSE NEAB Edition © Stanley Thornes (Publishers) Limited 1988

▌ Chemicals from air

1. Products; **2.** Reversible; **3.** Equilibrium; **4.** Reverse reaction; **5.** Ammonia; **6.** Left;

7. Ammonia; **8.** Right; **9.** Reverse reaction; **10.** Hydrogen; **11.** Nitrogen; **12.** Nitrogen;

13. Hydrogen; **14.** Catalyst; **15.** Iron; **16.** Increase; **17.** Slow; **18.** Liquefying;

19. Recycled; **20.** Exothermic; **21.** Oxygen; **22.** Platinum; **23.** Water; **24.** Nitrogen;

25. Ammonia; **26.** Nitric acid; **27.** Sulphuric acid; **28.** Quick acting; **29.** Slow acting;

30. Drinking water.

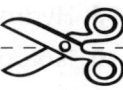

---

31. (a) (i) Increasing pressure increases the
                percentage of ammonia.   [1]

*Look at any curve. You will notice, at constant
temperature, the percentage of ammonia increases
with rise in pressure.*

     (ii) Increasing temperature decreases the
            percentage of ammonia.   [1]

     (iii) At 350 °C the establishment of
            equilibrium would be too slow. Raising
            the temperature decreases the yield of
            ammonia but speeds up the
            process.   [1]

  (b) (i) It speeds up both the forward reaction
                and the reverse reaction.   [1]

     (ii) The yield of ammonia remains
            unchanged.   [1]

*It is a very common mistake to believe that using a
catalyst will move the equilibrium to produce more
products.*

     (iii) Constant temperature   [1]
            Constant pressure   [1]
            No changes in amounts of reactants and
            products.   [1]

  (c) On cooling below −33 °C   [1]
        ammonia liquefies.   [1]
        Nitrogen and hydrogen are unchanged.   [1]

32. (a) Ammonium phosphate and potassium
            nitrate. *Both required.*   [1]

  (b) Nitric acid   [1]

  (c) urea + water → carbon dioxide + ammonia
                                                           [2]

1 mark for correct formulae and 1 mark for balancing.

Science: On Course for GCSE NEAB Edition Teacher's Book © Stanley Thornes (Publishers) Limited 1998

Atomic structure and bonding

1. Density; 2. Vibrating; 3. More; 4. Less regular; 5. Widely spaced; 6. Random;
7. Diffusion; 8. Gas; 9. Freezing; 10. Melting; 11. Evaporating; 12. Condensing;
13. Melting; 14. Gas; 15. Evaporating; 16. Boiling point; 17. Boil; 18. Atoms;
19. Mixture; 20. Magnet; 21. Combined; 22. Compound; 23. Sulphide;
24. Synthesis; 25. Neutron; 26. Electron; 27. Proton; 28. Nucleus; 29. Electrons;
30. Energy; 31. Positively charged; 32. Protons; 33. Electrons; 34. Electrons;
35. Hydrogen; 36. Proton; 37. Electrons; 38. Positively charged; 39. Isotopes;
40. Isotopes; 41. Electrons; 42. Atomic numbers; 43. Mass numbers; 44. Sodium;
45. Chlorine; 46. Neon; 47. Molecular; 48. Giant structure; 49. Regular; 50. High;
51. Low; 52. Metallic; 53. Covalent; 54. Ionic; 55. Iron; 56. Giant structure;
57. Giant structure; 58. Ionic; 59. Giant structure; 60. Covalent; 61. Bonding;
62. Sodium; 63. Chlorine; 64. Ions; 65. Lattice; 66. Electrostatic; 67. Molecule; 68. Pair;
69. Covalent; 70. Very weak; 71. Covalent; 72. Electrons; 73. Thermosoftening;
74. Cross-links; 75. Thermosetting.

76. (a)

state	shape	ease of compression	density
solid	regular	hard	high
liquid	**fills bottom of container**	**hard**	**high**
gas	**fills whole container**	**easy**	**low**

Allow ½ mark for each correct answer. Round up to nearest whole mark.   [3]

(b) It takes the shape of the bottom of the container.   [1]
It can be poured.   [1]

77.

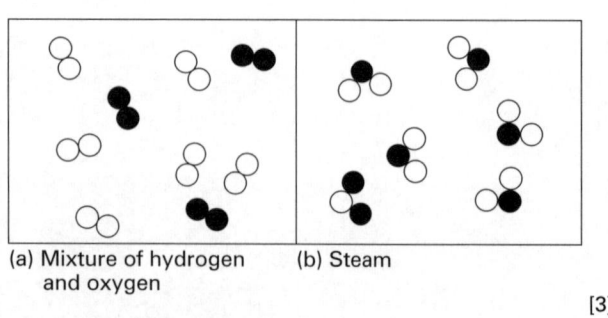

(a) Mixture of hydrogen   (b) Steam
and oxygen

[3]

1 mark for (a) and 2 marks for (b). For (b), there is 1 mark for showing hydrogen and oxygen atoms joined and 1 for showing two hydrogen atoms with each oxygen.

78. (a)

$$\times \times \quad \bullet \bullet$$
$$\times \, Cl \, \times \, Cl \, \bullet$$
$$\times \times \quad \bullet \bullet$$

Chlorine

(b)

$$H \quad O \quad H$$

Water

(c)

$$H \, Cl \,$$

Hydrogen chloride

(d)

$$O \, C \, O$$

Carbon dioxide   [4]

Science: On Course for GCSE NEAB Edition © Stanley Thornes (Publishers) Limited 1988

The use of 'dot and cross' diagrams shows which atoms the electrons came from. Remember, however, that all electrons are the same. Chlorine, water and hydrogen chloride contain single covalent bonds where one pair of electrons is shared. Carbon dioxide contains a double covalent bond where two pairs of electrons are shared. Another way of representing these molecules is

(a) Cl—Cl   (b)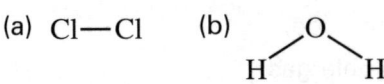

(c) H—Cl   (d) O═C═O

79. (a) **B** and **D**   [1]
Both **B** and **D** have good electrical
conductivity when solid.   [1]
(b) **C** and **E**   [1]
Low melting points and low boiling
points.   [1]

(c) **E**   [1]
Solution conducts electricity.   [1]

This question involves data handling. You are not expected to try to identify the five substances.

80. (a) NH_4Cl (s)   [1]

The correct formula and the state symbol are required for one mark.

(b) (i) Ammonia particles move faster than the
hydrogen chloride particles.   [1]
(ii) The ring forms closer to the pad soaked
in conc. hydrochloric acid.   [1]
(c) Particles are moving in all directions /
randomly.   [1]
Particles of nitrogen and oxygen fill the tube
and collide with the particles of ammonia
and hydrogen chloride, slowing their
movement along the tube.   [1]

Science: On Course for GCSE NEAB Edition Teacher's Book © Stanley Thornes (Publishers) Limited 1998

The Periodic Table

1. Helium; **2.** Inert; **3.** Chlorine; **4.** Metal; **5.** Alkaline; **6.** Acidic; **7.** Period; **8.** Group;

9. Atomic mass; **10.** Atomic number; **11.** Metals; **12.** Non-metals; **13.** Trends;

14. Metals; **15.** Alkali metals; **16.** Halogens; **17.** Noble gases; **18.** Transition metals;

19. Green; **20.** Paraffin oil; **21.** Shiny; **22.** Oxide; **23.** Hydrogen; **24.** Alkali; **25.** Liquid;

26. Salt; **27.** Iodine; **28.** Chlorine; **29.** Iodine; **30.** Chlorine; **31.** Silver chloride;

32. Hydrogen chloride; **33.** Sodium chloride; **34.** Chlorine; **35.** Sodium hydroxide;

36. Energy level; **37.** Lost; **38.** Gained; **39.** Alkali metal; **40.** Halogen; **41.** Noble gas;

42. Energy level; **43.** Electrons; **44.** Period; **45.** Group; **46.** Sodium chloride;

47. Sodium hydroxide; **48.** Chlorine; **49.** Sodium hydroxide; **50.** Acidic.

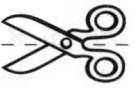

---

51. (a) As the atomic number increases (down the group) the melting and boiling points increase. [1]

(b) Rubidium [1]

The density is greater than the density of water (1.0 g/cm³).

(c) Rubidium is easily cut with a knife to give a shiny surface. [1]
Surface is silvery in colour. [1]
Surface rapidly tarnishes / goes dull on exposure to air. [1]

(d) (i) Violent reaction, hydrogen bursts into flame. [1]

Any answer will be accepted which implies a greater reactivity than potassium.

(ii) rubidium + water
$\rightarrow$ rubidium hydroxide + hydrogen [2]

1 mark for the correct formulae and 1 mark for balancing.

52. (a) **A** [1]

(b) (i) Fluorine is more reactive than chlorine so chlorine cannot displace fluorine. [1]

(ii) Displacement reaction [1]

(iii) $2KBr(aq) + Cl_2(g) \rightarrow 2KCl(aq) + Br_2(l)$ [2]

Predicting displacement reactions is important at Higher Tier.

53. (a) The potassium atom is larger than the sodium atom. [1]
The outer electron is less tightly held in potassium, so more easily lost. [1]
The electron arrangements are sodium 2, 8, 1 and potassium 2, 8, 8, 1. Both lose their outer electron to form an ion with a single positive charge.

(b) Bromine atom larger than chlorine atom. [1]
Chlorine better at attracting extra electron. [1]

Bromine and chlorine atoms have seven electrons in the outer energy level. They each attract an additional electron to form ions with a single negative charge.

(c) Argon has an electron arrangement of 2, 8, 8. [1]
This is a stable electron arrangement and it does not gain or lose electrons. [1]

(d) Going down group 2 there is an additional electron shell each time. [1]
Despite additional protons in the nucleus attracting electrons closer, the overall radius increases. [1]

(e) The electron arrangement in a potassium atom is 2, 8, 8, 1. [1]
Only a small amount of energy required to remove the outer electron. [1]
Removing a second electron would require an electron being lost from a full shell. [1]

Science: On Course for GCSE NEAB Edition © Stanley Thornes (Publishers) Limited 1988

Transferring energy

1. Temperatures; 2. Convection; 3. Radiation; 4. Radiation; 5. Thermal energy;

6. Particles; 7. Metals; 8. Gases; 9. Insulators; 10. Free electrons; 11. Kinetic;

12. Diffusion; 13. Gases; 14. Density; 15. Expands; 16. Electromagnetic radiation;

17. Infra-red; 18. Emitters; 19. Absorbers; 20. Infra-red; 21. Insulation; 22. Convection;

23. Conduction; 24. Convection; 25. Convection; 26. Conduction; 27. Conductors;

28. Loft insulation; 29. Insulator; 30. Reflects.

---

31. (a) Conduction [1]
    (b) Radiation [1]
    (c) Convection [1]

32. (a) Energy passes by conduction through the ceiling, [1]
    by convection through the roof-space, [1]
    and by conduction through the tiles. [1]

It is important to name the method involved when describing energy transfer by conduction, convection and radiation.

    (b) Air is trapped between the fibres. [1]
    This stops convection currents. [1]
    Energy can only pass through the insulation by conduction and air is a poor thermal conductor. [1]

33. (a) See graph. Correct plotting of all six points. [2]

1 mark for the correct plotting of three to five points.

        Best-fit straight line. [1]
    (b) (i) 42 J/s [1]
        (ii) 82 J/s [1]
        (iii) 165 J/s [1]
    (c) Yes, within the limits of precision of the graph. The energy flow at 10 °C (82 J/s) is twice that at 5 °C (42 J/s). [1]
    That at 20 °C (165 J/s) is twice that at 10 °C (82 J/s). [1]

No marks are awarded if the answer does not refer to the data from (b).

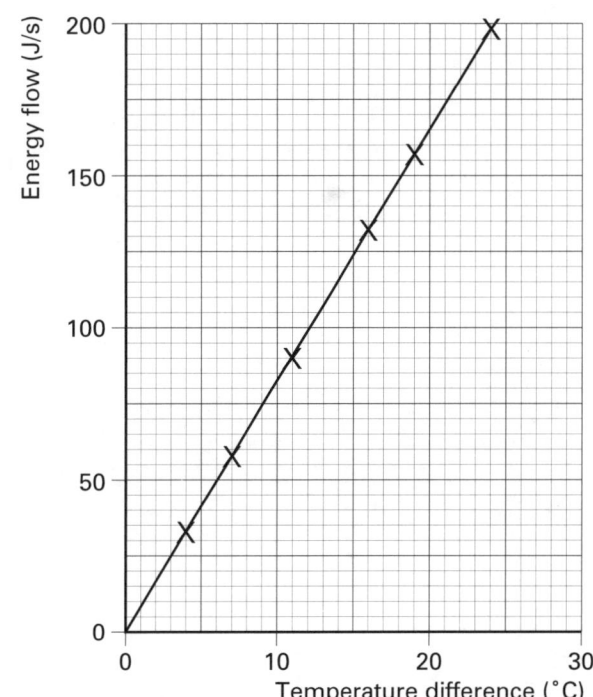

    (d) A straight line drawn through the origin with a shallower gradient. [1]
    The gradient is precisely half that of the original line. [1]

34. (a) Fish and chips:
    The paper prevents convection currents. [1]
    The layers of trapped air reduce the amount of energy conducted. [1]
    Burger:
    The lid prevents convection currents. [1]
    The pockets of trapped gas reduce the amount of energy conducted. [1]
    Chinese food:
    The container lid prevents convection currents. [1]

The foil container reduces the energy loss
by radiation. [1]
(b) Energy passes through the foil by
conduction. [1]

Convection currents in the air around the
container remove this energy. [1]
This could be reduced by placing the
container inside another insulating material
that has trapped air pockets. [1]

▎ Generating and using electricity

1. Energy; **2.** Heat; **3.** Movement; **4.** Light; **5.** Sound; **6.** Heat; **7.** Movement; **8.** Sound;

9. Light; **10.** Sound; **11.** Heat; **12.** Heat; **13.** Electricity; **14.** Efficiency; **15.** Light;

16. Electricity; **17.** Power; **18.** Kilowatt-hours; **19.** Watts; **20.** Fossil fuels; **21.** Oil;

22. Non-renewable; **23.** Sun; **24.** Coal; **25.** Pressure; **26.** Turbines; **27.** Generator;

28. Electricity; **29.** Temperature; **30.** Radioactive; **31.** Radioactive; **32.** Geothermal;

33. Electricity; **34.** Turbine; **35.** Renewable; **36.** Waves; **37.** Generators;

38. Hydroelectric; **39.** Atmosphere; **40.** Noise; **41.** Renewable; **42.** Electricity;

43. Batteries; **44.** Mains electricity; **45.** Power; **46.** Sun; **47.** Efficiency;

48. Gravitational potential; **49.** Kinetic; **50.** Turbines.

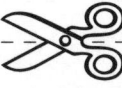

---

51. (a) Electricity [1]
    (b) Toaster and hairdryer [2]

Award 1 mark for each.

    (c) Vacuum cleaner and hairdryer [2]

Award 1 mark for each.

    (d) (i) It is more powerful/brighter/cheaper to operate/the mains does not run out. [1]
       (ii) It is portable/does not need a connecting wire. [1]

52. (a) Energy from electricity [1]
      is transferred to gravitational potential energy of the water. [1]
    (b) The water falls down pipes, transferring gravitational potential energy to kinetic energy. [1]
      It then passes through turbines, losing kinetic energy. [1]
      The turbines turn generators that generate electricity. [1]

    (c) 80% of the original energy from electricity [1]
      is returned to electricity. [1]
      Some energy is transferred to heat in the moving machinery, [1]
      and the water still has some kinetic energy as it leaves the turbines. [1]

53. (a) efficiency = 40 ÷ 100 = 0.4 [1]

Efficiency can also be expressed as a percentage; in this case it is 40%.

    (b) energy input = 2000 MJ ÷ 0.4 [1]
                   = 5000 MJ [1]
    (c) 3000 MJ [1]
    (d) In the steam from the cooling towers [1]
      In the waste gases from the chimney [1]
      In the cooling water [1]
    (e) It could be used for space heating of homes, workplaces or greenhouses. [1]
      It could be used to provide a supply of hot water for homes and workplaces. [1]

Award marks for other possible uses of this energy.

Science: On Course for GCSE NEAB Edition Teacher's Book © Stanley Thornes (Publishers) Limited 1998

Current, charge and circuits

1. Forces; **2.** Repel; **3.** Attract; **4.** Electrons; **5.** Friction; **6.** Negatively; **7.** Positively;

8. Static; **9.** Voltage; **10.** Conductor; **11.** Static; **12.** Current; **13.** Amps; **14.** Ammeter;

15. Electrons; **16.** Negative; **17.** Positive; **18.** Dissolved; **19.** Negative; **20.** Positive;

21. Negative; **22.** Current; **23.** Time; **24.** Series; **25.** Parallel; **26.** Resistance;

27. Voltage; **28.** Voltage; **29.** Current; **30.** Components; **31.** Battery; **32.** Closed switch;

33. Diode; **34.** Resistor; **35.** Variable resistor; **36.** Ammeter; **37.** Voltmeter; **38.** Lamp;

39. Energy; **40.** Voltmeter; **41.** Parallel; **42.** Voltage; **43.** Increase; **44.** Decrease;

45. Voltmeter; **46.** Ohms; **47.** Increase; **48.** Current; **49.** Decreases; **50.** Decreases.

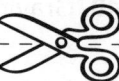

---

51. (a) Negative   [1]
     (b) Positive   [1]
     (c) The comb and the hair have opposite charges.   [1]
       They attract each other.   [1]

52. A1: 1.5 A   [1]
     A2: 1.5 A   [1]
     A3: 2.3 A   [1]
     A4: 1.6 A   [1]
     A5: 1.1 A   [1]
     A6: 1.6 A   [1]

A common misconception is that electric current is used up to create heat and light. This is not the case; the current has to be the same at all points in a series circuit and the total current passing into a junction in a parallel circuit has to equal the current passing out of the junction.

53. (a)

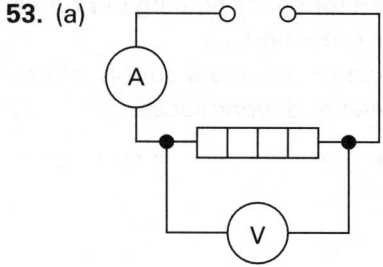

Award 1 mark for each meter labelled correctly.   [2]

     (b) Variable resistor   [1]
       Correct symbol: [1]

     placed either side of the heater, between the power supply and the connection to the voltmeter.   [1]

Award 1 mark for getting the symbol correct even if it is in the wrong position.

     (c)   [3]

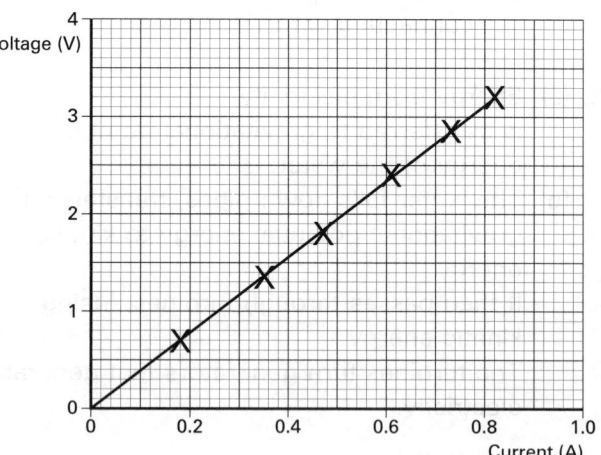

2 marks for plotting all the points correctly, 1 mark for plotting between three and five correctly, 1 mark for drawing the best-fit straight line.

Science: On Course for GCSE NEAB Edition © Stanley Thornes (Publishers) Limited 1988

(d)   [2]

current (A)	voltage (V)
0.20	0.8
0.40	1.55
0.60	2.35
0.80	3.15

2 marks for completing the table correctly from the graph.

(e) Yes, since doubling the current from 0.2 A to 0.4 A causes the voltage to double, within the limits of precision of the graph.   [1]
The voltage also doubles when the current is doubled from 0.4 A to 0.8 A.   [1]

No marks if the answer does not use the data contained in (d).

(f) The voltage is 1.95 V   [1]

$$\text{resistance} = \frac{\text{voltage}}{\text{current}} = \frac{1.95\,\text{V}}{0.5\,\text{A}}$$ [1]

$$= 3.9\,\Omega$$ [1]

The unit must be correct for all 3 marks to be awarded. The correct numerical answer with a wrong or missing unit gains 2 marks.

Science: On Course for GCSE NEAB Edition Teacher's Book © Stanley Thornes (Publishers) Limited 1998

Using electricity

1. Direct; 2. Alternating; 3. Live; 4. Neutral; 5. Earth; 6. Live; 7. Neutral; 8. Earth; 9. Earth; 10. Insulation; 11. Neutral; 12. Live; 13. Earth; 14. Neutral; 15. Earth; 16. Fuse; 17. Fuse; 18. Current; 19. Earth; 20. Resistance; 21. Fuse; 22. Conductors; 23. Insulated; 24. Earth; 25. Circuit breaker; 26. Fuses; 27. Energy; 28. Watts; 29. Voltage; 30. Coulomb.

---

31. (a) Earth   [1]
    (b) Neutral   [1]
    (c) Live   [1]
    (d) Live   [1]

32. (a) Electrocution   [1]
    (b) Fuse   [1]
    (c) A large current passes to earth.   [1]
         This causes the fuse to melt.   [1]

A large current passes because there is a complete circuit from live to earth with very little resistance.

    (d) The casing is plastic/an insulator,   [1]
         so it cannot become live.   [1]

33. (a)   [2]

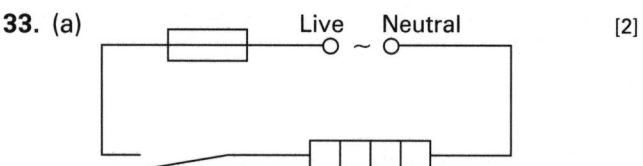

1 mark each for placing the switch and the fuse in the live conductor.

    (b) (i) Earth   [1]

As the earth wire is not normally part of the circuit, it is usually omitted from circuit diagrams that show the current path.

         (ii) To the metal casing   [1]

    (c) power = current × voltage   [1]
              = 4.5 A × 230 V   [1]
              = 1035 W   [1]

34. (a) It is lower than when inside.   [1]
    (b) There is good contact between a person's
         feet and the ground.   [1]

An alternative correct answer is that when a person is inside, the flooring insulates the person from the ground.

    (c) Water is a good conductor of mains
         electricity.   [1]
         Condensation could form a circuit from the
         live pin.   [1]
         OR water on the body creates a low-
         resistance path to earth.   [1]

Although water is not a good conductor of low-voltage electricity, the mains voltage is high enough to ionise the water so that it conducts.

    (d) It should prevent death,   [1]
         but it would cause muscular spasms   [1]
         and other serious injury if all the residual
         current passed through the body.   [1]

Science: On Course for GCSE NEAB Edition © Stanley Thornes (Publishers) Limited 1988

Magnetism and electromagnetism

1. Attract; 2. Repel; 3. Poles; 4. North; 5. South; 6. Repel; 7. Attract; 8. Magnetic field;

9. Electromagnets; 10. Current; 11. Coil; 12. Bar magnet; 13. Iron; 14. Fixed magnet;

15. Paper cone; 16. Coil; 17. Armature; 18. Iron core; 19. Switch contacts; 20. Coil;

21. Current; 22. Alternating; 23. Frequency; 24. Current; 25. Armature; 26. Attracted;

27. Electromagnetism; 28. Magnetic field; 29. Opposite directions; 30. Magnetic field;

31. Electromagnetic induction; 32. Ammeter; 33. Current; 34. Magnet; 35. Speed;

36. Magnet; 37. Electromagnet; 38. Generator; 39. Speed; 40. Magnetic field; 41. Coil;

42. Coil; 43. Slip rings; 44. Carbon brushes; 45. Voltage; 46. Current; 47. Voltage;

48. Voltage; 49. Primary; 50. Secondary.

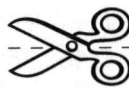

---

51. (a) A and D [2]

1 mark for each correct response.

    (b) B and C [2]

1 mark for each correct response.

52. (a) The current/voltage to the relay is switched
on, [1]
magnetising the relay core. [1]
The L-shaped armature is attracted to the
core, pressing the switch contacts
together. [1]

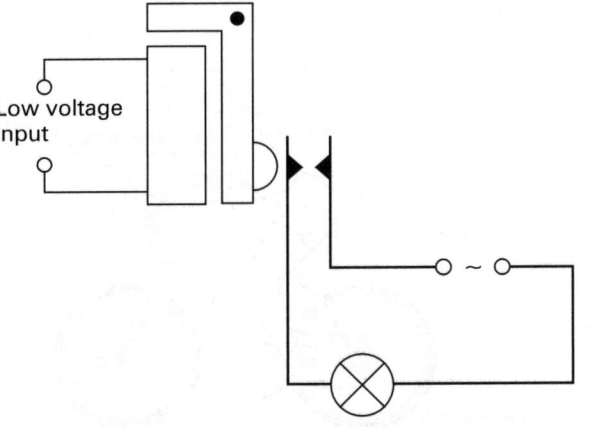

Low voltage
input

    (b) See diagram. The completed diagram
should show a mains supply in series
with the switches *(1 mark)* and a lamp
(1 mark). [2]

    (c) The relay enables the low-voltage control
circuit to switch [1]
the mains-operated lamp. [1]

*It is important to emphasise that there are two
separate circuits here; the circuit containing the
relay coil is low-voltage and carries a low current.
The relay contacts form part of a high-voltage,
higher-current circuit.*

53. (a) (i) The force reverses. [1]
       (ii) The force reverses. [1]
    (b) Increasing the voltage or current. [1]
       Increasing the magnetic field strength. [1]

*'Increasing the current and increasing the voltage'
does not gain 2 marks, since one is a consequence of
the other.*

54. (a) The current in the left-hand coil has a
magnetic field. [1]
This passes through the right-hand coil. [1]
A current is induced because of the
changing (increasing) magnetic field. [1]

*When answering questions about electromagnetic
induction, always refer to any **changes** of magnetic
field that occur.*

    (b) When the magnetic field is steady there is
no further change, [1]
so there is no induced current. [1]
    (c) The pointer moves, [1]
then returns to zero. [1]
The change in magnetic field is in the
opposite direction, so the induced current
is in the opposite direction. [1]

Science: On Course for GCSE NEAB Edition Teacher's Book © Stanley Thornes (Publishers) Limited 1998

Force and motion

1. Speed; **2.** Gradient; **3.** Distance; **4.** Speed; **5.** Velocity; **6.** Distance; **7.** Velocity;
8. Accelerating; **9.** Velocity; **10.** Acceleration; **11.** Speed; **12.** Force; **13.** Equal;
14. Opposite; **15.** Friction; **16.** Resistive; **17.** Braking distance; **18.** Speed; **19.** Mass;
20. Stopping distance; **21.** Thinking distance; **22.** Reaction time; **23.** Weight;
24. Weight; **25.** Air resistance; **26.** Weight; **27.** Air resistance; **28.** Terminal velocity;
29. Joules; **30.** Kinetic.

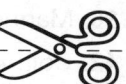

---

31. (a) 600 s [1]
(b) 150 s [1]

The person was resting between the times 200 s and 350 s shown on the graph.

(c) C [1]
   The line is steepest/has the greatest gradient or slope. [1]

(d) speed = $\dfrac{\text{distance travelled}}{\text{time taken}}$ [1]

   $= \dfrac{400\,\text{m}}{200\,\text{s}}$ [1]

   $= 2.0\,\text{m/s}$ [1]

(e) average speed $= \dfrac{1000\,\text{m}}{600\,\text{s}}$ [1]

   $= 1.7\,\text{m/s}$ [1]

(f) At times the person was walking faster and at other times slower. [1]

The person did not maintain a steady speed overall. There are three different speeds (including zero) shown on the graph, none of which is the same as the average speed.

32. (a) [3]

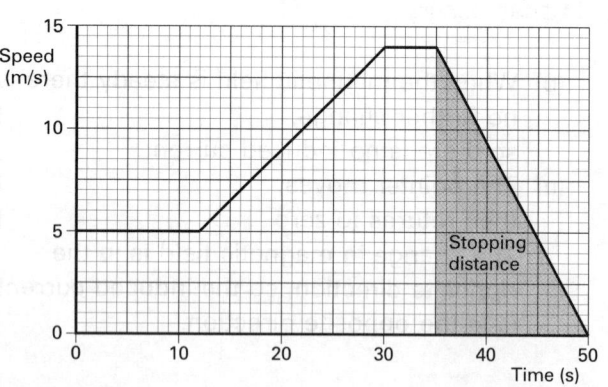

Deduct 1 mark for each mistake made.

(b) distance = average speed × time [1]
   = 9.5 m/s × 18 s [1]
   = 171 m [1]

Alternatively this can be calculated as the area between the graph line and the time axis during the period 12 s to 30 s.

(c) distance = (5 m/s × 12 s) + (14 m/s × 5 s) [1]
   = 60 m + 70 m = 130 m [1]

(d) See graph. [1]

(e) *Any two from:*
   condition of brakes, condition of road, condition of tyres,
   mass of cyclist, speed of cyclist [2]
   1 mark for each.

(f) (i) [2]

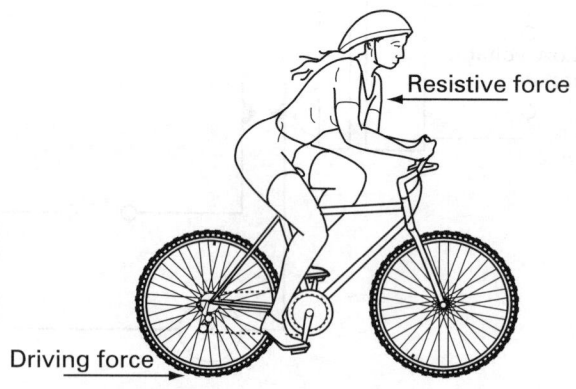

Resistive force

Driving force

1 mark for each correctly labelled force.

(ii) The cyclist is travelling at a constant speed. [1]
   The resistive force and driving force are equal/the forces are balanced. [1]

Science: On Course for GCSE NEAB Edition © Stanley Thornes (Publishers) Limited 1988

(iii) [2]

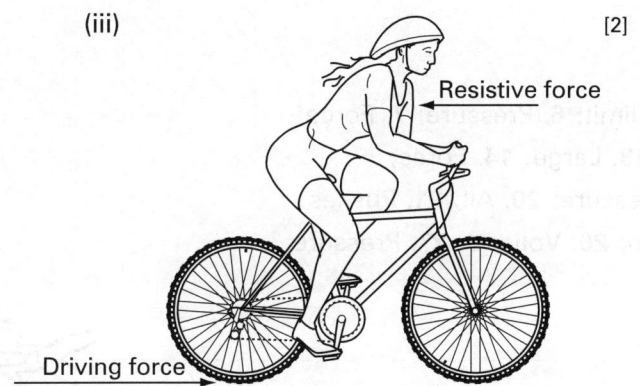

1 mark for showing both forces in the correct directions and 1 mark for the driving force arrow being bigger than the resistive force arrow.

33. (a) C [2]

(b) acceleration = $\dfrac{\text{increase in velocity}}{\text{time taken}}$ [1]

$= \dfrac{14\,\text{m/s}}{10\,\text{s}}$ [1]

$= 1.4\,\text{m/s}^2$ [1]

(c) force = mass × acceleration [1]

$= 925\,\text{kg} \times 1.4\,\text{m/s}^2$ [1]

$= 1295\,\text{N}$ [1]

(d) The deceleration is reduced. [1]

It takes longer for the car to stop. [1]

The braking distance is increased. [1]

The driver should leave a greater than normal distance between the car and the vehicle travelling in front. [1]

Forces and their effects

1. Elastic; 2. Elastic; 3. Force; 4. Proportional; 5. Elastic limit; 6. Pressure; 7. Force;
8. Pascal; 9. Large; 10. Small; 11. Pressure; 12. Small; 13. Large; 14. Force;
15. Pressure; 16. Force; 17. Hydraulic; 18. Force; 19. Pressure; 20. All; 21. Pushes;
22. Proportional; 23. Pressure; 24. Random; 25. Collision; 26. Volume; 27. Pressure;
28. Increase; 29. Halves; 30. Proportional.

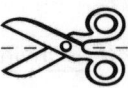

31. (a) Wooden bookshelf, pencil eraser,
     sponge   [2]

1 mark for two of the three correct.

  (b) (i) See table.   [3]

Deduct 1 mark for each error.   [3]

  (ii)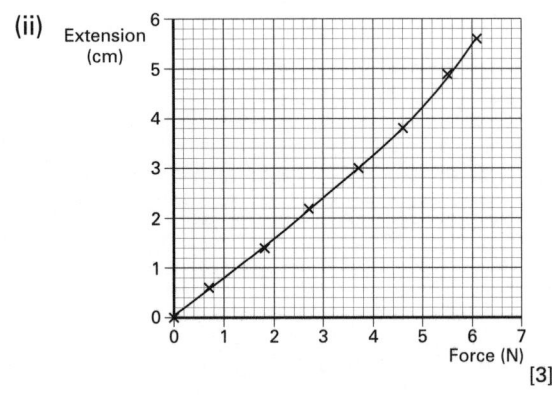

Extension (cm) vs Force (N)

[3]

*Award 2 marks for plotting all the points correctly
(1 mark for plotting four to seven points correctly)
and 1 mark for the best-fit line.*

  (iii) Increasing the force causes an increase
        in extension.   [1]
        The increase in extension is uniform for
        small forces,   [1]
        and is non-uniform for larger forces. [1]
  (iv) For forces up to 3.7 N.   [1]
        The graph is a straight line through the
        origin.   [1]
  (v)   For forces up to 4.6 N.   [1]
        The spring returns to its original size
        when the force is removed.   [1]

force (N)	length of spring (cm)	length when load removed (cm)	total extension (cm)
0	4.6	4.6	0
0.7	5.2	4.6	0.6
1.8	6.0	4.6	1.4
2.7	6.8	4.6	2.2
3.7	7.6	4.6	3.0
4.6	8.4	4.6	3.8
5.5	9.5	4.8	4.9
6.1	10.2	5.1	5.6

Science: On Course for GCSE NEAB Edition © Stanley Thornes (Publishers) Limited 1988

32. (a) pressure $= \dfrac{\text{force}}{\text{area}}$ [1]

$= \dfrac{200\,\text{N}}{0.25\,\text{m}^2}$ [1]

$= 800\,\text{Pa}$ [1]

1 Pa (pascal) is the same as 1 N/m².

(b) 800 Pa [1]

(c) force = pressure × area [1]

= 800 Pa × 3.6 m² [1]

= 2880 N [1]

(d) Using hydraulics enables forces to be multiplied easily. [1]
Liquids transmit pressure equally in all directions, so the pressure can be transmitted round corners and to places difficult to access in other ways. [1]

33. (a) Gas pressure is due to collisions [1] between the particles and the walls of the container. [1]

It is a common misconception to describe gas pressure as being due to collisions between gas particles. Pressure is exerted on the container walls and other objects that the gas is in contact with.

(b) $P_1 \times V_1 = P_2 \times V_2$ [1]

$95 \times 0.8 = 1 \times V_2$ [1]

$V_2 = 76$ litres [1]

(c) 75.2 litres [1]

Gas flows out of the container until the pressure in the container is the same as atmospheric pressure, so the container is left full of gas at atmospheric pressure.

Science: On Course for GCSE NEAB Edition Teacher's Book © Stanley Thornes (Publishers) Limited 1998

The Earth and beyond

1. Solar System; **2.** Star; **3.** Galaxy; **4.** Planets; **5.** Sun; **6.** Gravitational; **7.** Orbital;
8. Year; **9.** Constellations; **10.** Sun; **11.** Moon; **12.** Satellites; **13.** Orbit; **14.** Orbit;
15. Geostationary; **16.** Sun; **17.** Comets; **18.** Gravitational; **19.** Energy; **20.** Stars;
21. Fusion; **22.** Gravitational; **23.** Red giant; **24.** White dwarfs; **25.** Fusion;
26. Supernova; **27.** Earlier star; **28.** Galaxies; **29.** Big Bang; **30.** Universe.

---

31. (a) Sun   [1]
     (b) Sun   [1]
     (c) Planet   [1]
     (d) Moon *(1 mark)* and satellite *(1 mark)*   [2]
     (e) Galaxy   [1]

32. (a) Mercury and Venus   [2]

*1 mark each. The table does not give orbital speeds,
but the closer a planet is to the Sun, the faster it
moves in its orbit.*

     (b) Venus   [1]

*As a planet rotates once on its own axis it
completes one day, that is one cycle of daylight and
night.*

     (c) Mercury, Mars, Venus, Earth, Jupiter   [2]

1 mark if one or two are out of order.

     (d) The sketch should show: Mercury the
          nearest planet to the Sun, followed by
          Venus, Earth, Mars and Jupiter.   [2]

1 mark if one planet is in the wrong position.

     (e) Mars has further to travel in its orbit.   [1]
          The orbital speed of Mars is less than that of
          the Earth.   [1]
     (f) All points plotted correctly.   [2]

1 mark if three or four are plotted correctly.

Smooth curve drawn.   [1]

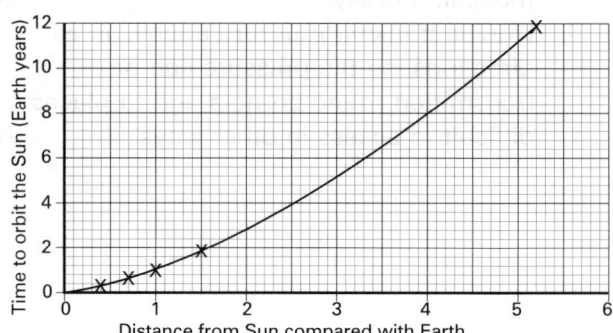

     (g) The orbit time increases   [1]
          with increasing distance from the Sun.   [1]

*It is important to state 'with increasing distance
from the Sun' as 'with distance from the Sun' is
ambiguous, it could mean increasing or decreasing.*

     (h) Between 2.1 and 3.4 times the distance of
          the Earth from the Sun.   [1]
     (i) Mars and Jupiter   [1]

33. (a) Parts of the cloud contracted owing to
          gravitational forces.   [1]
          This contraction caused heating.   [1]
          When it was hot enough it started to give
          out light.   [1]
     (b) The fusion of nuclei *(1 mark)*: hydrogen
          nuclei to helium nuclei *(1 mark)*.   [2]
     (c) The Sun will cool and expand.
          It will form a red giant.
          It will then contract to a white dwarf.
          It will become dimmer as further cooling
          occurs.   [3]

*1 mark for each of any three points. Stars cool as
they expand and heat as they contract.*

Science: On Course for GCSE NEAB Edition © Stanley Thornes (Publishers) Limited 1988

▌ Wave properties

1. Energy; **2.** Vibration; **3.** Oscillation; **4.** Cycle; **5.** Wavelength; **6.** Frequency;

7. Amplitude; **8.** Hertz; **9.** Refraction; **10.** Reflected; **11.** Reflection; **12.** Refraction;

13. Diffraction; **14.** Wavelength; **15.** Gases; **16.** Longitudinal; **17.** Parallel;

18. Frequency; **19.** Amplitude; **20.** Light; **21.** Vacuum; **22.** Echoes; **23.** Ultrasound;

24. Reflection; **25.** Speed; **26.** Ultrasound; **27.** Reflections; **28.** Frequencies;

29. Wavelengths; **30.** Diffracted; **31.** Longitudinal; **32.** Transverse; **33.** Vibrations;

34. Sound; **35.** Longitudinal; **36.** Light; **37.** Transverse; **38.** Frequency; **39.** Transverse;

40. Primary; **41.** Secondary; **42.** Longitudinal; **43.** Transverse; **44.** Primary;

45. Secondary; **46.** Seismometers; **47.** Primary; **48.** Mantle; **49.** Core; **50.** Primary.

---

51. (a) (i) Four   [1]

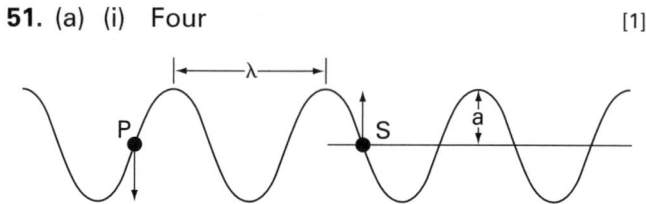

    (ii) The distance should show one wave cycle, i.e. a crest and a trough.   [1]

    (iii) The distance should extend from the centre to the maximum upward or downward displacement.   [1]

    (iv) P is down the page.   [1]
          S is up the page.   [1]

The wave is moving from left to right, so the parts of the waveform to the left of P and S are about to pass through them.

  (b) 2 Hz   [1]

The frequency is the number of waves that pass a given point in one second, so it can be calculated from number of wave cycles ÷ time.

52. (a)   [1]

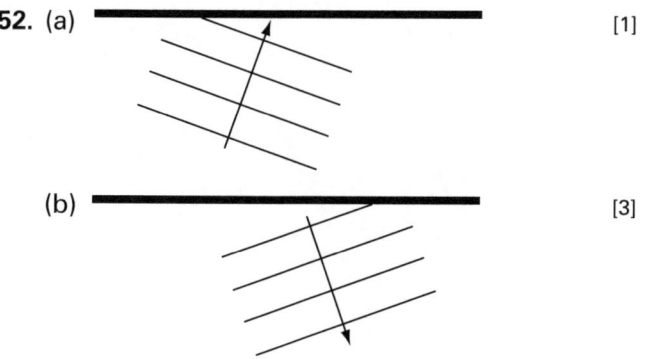

  (b)   [3]

1 mark for the wavefronts in the correct direction, 1 mark for the same spacing as before reflection and 1 mark for the correct direction of travel.

  (c) (i)   Stays the same.   [1]
    (ii) Stays the same.   [1]
    (iii) Stays the same.   [1]

53. (a)   [4]

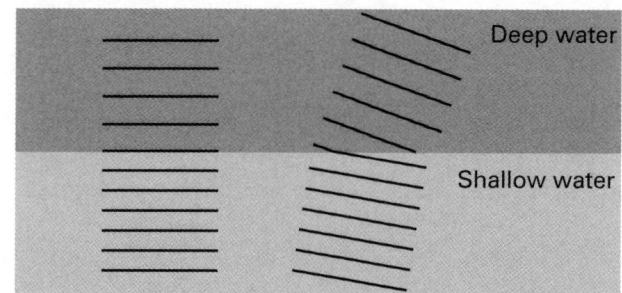

For each diagram, 1 mark for the correct reduction in wavelength and 1 mark for the correct orientation of the wavefronts.

  (b) (i)   Decreases.   [1]
    (ii) Stays the same.   [1]
    (iii) Decreases.   [1]

When waves are refracted the speed decreases or increases; this causes an equivalent change in the wavelength but the frequency is unchanged.

Science: On Course for GCSE NEAB Edition Teacher's Book © Stanley Thornes (Publishers) Limited 1998

54. (a) Diffraction [1]

(b) The wavelength of the waves. [1]

1 mark for each. Note that frequency is not an alternative to wavelength. 3 cm radio waves and 3 cm water waves are diffracted in a similar way when they pass through similar size openings but their frequencies are very different.

(c) The left-hand diagram should show semicircular waves that have spread into the shadow region. [1]
The right-hand diagram should show some spreading, but less than in the left-hand diagram. [1]

(d) speed = wavelength × frequency [1]
= 12.5 m × 0.3 Hz [1]
= 3.75 m/s [1]

55. (a) [1]

(b) The hand moves backwards and forwards/ vibrates/oscillates [1]
parallel to the spring. [1]

56. (a) (i) The same [1]
(ii) The same [1]
(iii) Smaller [1]

(b) The amplitude *(1 mark)* increases *(1 mark)*. [2]

(c) The frequency *(1 mark)* decreases *(1 mark)*. [2]

In parts (b) and (c) there is 1 mark for correctly identifying the wave property that changes and 1 mark for the correct change. Award these marks independently.

Electromagnetic radiation

1. Wavelength; **2.** Sound; **3.** Mirrors; **4.** Reflection; **5.** Critical angle;

6. Total internal reflection; **7.** Wavelength; **8.** Refraction; **9.** Dispersion; **10.** Spectrum;

11. Wavelength; **12.** Electromagnetic spectrum; **13.** Vacuum; **14.** Frequency;

15. Radio waves; **16.** Wavelength; **17.** Frequency; **18.** Gamma rays; **19.** Frequency;

20. Wavelength; **21.** Radio waves; **22.** Diffracted; **23.** Microwaves; **24.** Infra-red;

25. Light; **26.** Total internal reflection; **27.** Prisms; **28.** Ultraviolet; **29.** Gamma rays;

30. Gamma rays.

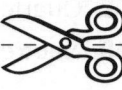

- -

31. (a) Light hits the boundary at an angle greater than the critical angle. [1]
So it is all reflected. [1]
(b) Two fibres/bundles of fibres are passed down the patient's throat into the wind-pipe. [1]
Light is shone down one fibre, lighting up the wind-pipe. [1]
The reflected light passes up the second fibre to a camera. [1]

32. (a) The wavefronts spread into the shadow region. [1]
and are the same wavelength as those to the left of the barrier. [1]
(b) Diffraction. [1]
(c) Light has a much shorter wavelength, so less spreading occurs. [1]
The light passes through as a straight beam. [1]

Diffraction effects depend on the wavelength, the greater the wavelength, the more spreading occurs.

33. (a) *Any two from:*
Travel at the same speed in a vacuum
All transverse
All electromagnetic
Can all travel in a vacuum

1 mark for each. [2]

(b) gamma rays, ultraviolet, light, infra-red, microwaves, radio waves [3]
Deduct 1 mark for each pair out of sequence.

(c) Microwaves and infra-red [2]
1 mark for each.

(d) (i) X-rays pass through skin and flesh. [1]
They are absorbed by bone. [1]
(ii) X-rays can damage cells/chromosomes. [1]
The operators should be in a separate room when the equipment is being operated [1]
or protect important body organs with lead shielding. [1]

34. (a) Left-hand diagram: light passes through first face undeviated. [1]
Internal reflection (through 90°) and out through second face undeviated. [1]
Right-hand diagram: correct internal reflection. [1]
Correct second reflection and out through second face undeviated. [1]
(b) Any correct use, e.g. camera, binoculars, prism periscope. [1]
(c) Suitable arrangement of light beam and prism. [1]
1 mark for each correct refraction. [2]
Blue deviated more than red. [1]

Radioactivity

1. Nucleus; **2.** Alpha; **3.** Beta; **4.** Gamma; **5.** Protons; **6.** Electrons;

7. Electromagnetic radiation; **8.** Background radiation; **9.** Electrons; **10.** Radiation;

11. Beta; **12.** Gamma; **13.** Beta; **14.** Gamma; **15.** Positively; **16.** Alpha; **17.** Nucleus;

18. Positive; **19.** Electrons; **20.** Nucleus; **21.** Neutrons; **22.** Mass; **23.** Positive;

24. Nucleus; **25.** Proton; **26.** Negative; **27.** Electrons; **28.** Beta; **29.** Neutrons;

30. Nucleons; **31.** Neutrons **32.** Isotopes; **33.** Protons; **34.** Neutrons; **35.** Radionuclide;

36. Proton; **37.** Decreases; **38.** Radioactivity; **39.** Random; **40.** Half-life; **41.** Counts/s;

42. Quarter; **43.** Gamma; **44.** Penetration; **45.** Alpha; **46.** Ionisation; **47.** Fission;

48. Neutron; **49.** Radioactive; **50.** Neutrons.

51. (a) Gamma [1]
    (b) Alpha [1]
    (c) Alpha and beta [1]
    (d) Alpha, beta and gamma [1]

52. (a) From space/the Sun. [1]
    (b) It is absorbed by the atmosphere. [1]
    (c) People who spend a lot of time flying/in space. [1]

53. (a) A Geiger–Müller tube or photographic film. [1]
    (b) It emits gamma radiation which penetrates soil.
    It is soluble in water.
    It has a half-life long enough for it to be detected after it has travelled through the pipes.
    Its half-life is short enough for it to decay to a low level within a few days.

    Award 1 mark for each point up to 3 marks. [3]
    When choosing a radioactive isotope to do a particular job, the half-life and emission of the isotope are both important to consider.

    (c) Keep as great a distance as possible from the pipeline and any leaking water.
    Monitor the amount of radiation that the body is subjected to.
    Wear protective clothing.

    Award 1 mark for each up to 2 marks. [2]

54. (a) The background radiation is measured over a period of time, e.g. 5 minutes. [1]
    The total count is divided by the time in seconds to give the average count/s. [1]
    This is then subtracted from each reading. [1]

    (b)

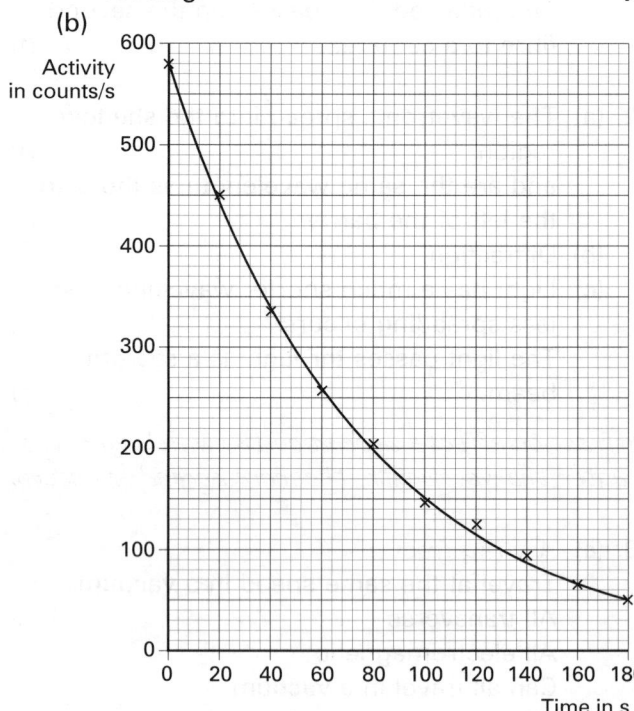

    Correct plotting of points [2]
    Smooth curve drawn [1]

    2 marks for plotting all ten points correctly.
    1 mark for plotting five to nine points correctly.

(c) Radioactive decay is a random process so it
   never fits a precise pattern. [1]
(d) One reading (approximately 50 s) taken
   from graph. [1]
   Two or more readings taken. [1]
   An average value calculated. [1]

Because of the random nature of radioactive decay there may be a slight variation in the time taken for the activity to reduce to half of its original value. This is why it is important to work out an average value.

Formulae, equations and calculations

1. Reactants [1]

2. Products [1]

3. **Magnesium oxide + sulphuric acid →
   magnesium sulphate + water** [2]

1 mark for two or three correct names.

4. **Barium nitrate + sulphuric acid →
   barium sulphate + nitric acid** [2]

*2 marks for three correct names added and 1 mark
for two.*

5.

formula	name
NaCl	sodium chloride
Na_2CO_3	sodium carbonate
Na_2SO_4	sodium sulphate
$NaNO_3$	sodium nitrate
$Ca(OH)_2$	calcium hydroxide
$Ca(HCO_3)_2$	calcium hydrogencarbonate
$MgCl_2$	magnesium chloride
$MgSO_4$	magnesium sulphate
$Mg(NO_3)_2$	magnesium nitrate
$Al_2(SO_4)_3$	aluminium sulphate
H_2SO_4	sulphuric acid
HCl	hydrochloric acid
$ZnCl_2$	zinc chloride
$Zn(OH)_2$	zinc hydroxide
$ZnCO_3$	zinc carbonate
PbO	lead(II) oxide
$Pb(NO_3)_2$	lead(II) nitrate
$Pb(OH)_2$	lead(II) hydroxide
$PbCl_2$	lead(II) chloride
$PbCO_3$	lead(II) carbonate

[10]

$\frac{1}{2}$ *mark for each answer.*

6. (a) $FeSO_4$ (b) $Fe_2(SO_4)_3$
   (c) $Fe(OH)_2$ (d) Fe_2O_3 [4]

7. (a) $2Mg(s) + O_2(g) \rightarrow 2MgO(s)$ [2]
   (b) $2Na(s) + 2H_2O(l) \rightarrow 2NaOH(aq) + H_2(g)$ [2]
   (c) $2NaHCO_3(s) \rightarrow$
   $$Na_2CO_3(s) + H_2O(l) + CO_2(g)$$ [2]
   (d) $CH_4(g) + 2O_2(g) \rightarrow CO_2(g) + 2H_2O(l)$ [2]
   (e) $CaO(s) + 2HNO_3(aq) \rightarrow$
   $$Ca(NO_3)_2(aq) + H_2O(l)$$ [2]
   (f) $MgCO_3(s) + 2HCl(aq) \rightarrow$
   $$MgCl_2(aq) + H_2O(l) + CO_2(g)$$ [2]
   (g) $4NH_3(g) + 3O_2(g) \rightarrow 2N_2(g) + 6H_2O(g)$ [2]

*Allow 2 marks for each equation if completely
correct and 1 mark for a slight mistake.*

8. Solid, liquid, gas, aqueous solution
   (solution in water). [4]

9. (a) $Zn(s) + H_2SO_4(aq) \rightarrow ZnSO_4(aq) + H_2(g)$ [2]
   (b) $2Al_2O_3(l) \rightarrow 4Al(l) + 3O_2(g)$ [2]
   (c) $Ca(OH)_2(aq) + CO_2(g) \rightarrow CaCO_3(s) + H_2O(l)$ [2]

*Allow 2 marks for each equation if completely
correct and 1 mark for a slight mistake.*

10. (a) $2Cl^- \rightarrow Cl_2 + 2e^-$ [1]
    (b) $Zn \rightarrow Zn^{2+} + 2e^-$ [1]
    (c) $O^{2-} + 2H^+ \rightarrow H_2O$ [1]

11. (a) 4 [1]
    (b) 10 [1]
    (c) 64 [1]
    (d) 56 g of iron react with 32 g of sulphur, so
    $(56 \div 32) = 1.75$ g of iron are needed for
    each gram of sulphur. [1]

Science: On Course for GCSE NEAB Edition © Stanley Thornes (Publishers) Limited 1988

12. (a) Calcium carbonate: 100 g   [1]

Calcium oxide: 56 g   [1]

(b) 5   [1]

Divide the mass of calcium carbonate by the formula mass of calcium carbonate.

(c) 5   [1]

From the equation, 1 formula mass of calcium carbonate produces 1 formula mass of calcium oxide and 1 formula mass of carbon dioxide.

(d) 120 dm^3 [2]

13. (a) 3.60 g   [1]

Subtract the mass of the combustion boat from the mass of the combustion boat and contents before reduction.

(b) 3.20 g   [1]

Subtract the mass of the combustion boat from the mass of the combustion boat and contents after reduction.

(c) $\dfrac{3.20}{64} = 0.05$ [1]

Calculate the number of formula masses by dividing the mass by the relative atomic mass of copper.

(d) 0.40 g   [1]

Subtract the mass of the combustion boat and contents after reduction from the mass of the combustion boat and contents before reduction.

(e) $\dfrac{0.40}{16} = 0.025$ [1]

(f) Cu_2O [1]

There are twice as many copper atoms as oxygen atoms. The simplest formula is Cu_2O. Don't be put off by the fact that the black copper oxide you have probably seen is CuO. This copper oxide is the less the common copper(I) oxide which is red in colour.

(g) If copper is hot when the apparatus is taken apart, copper will react with oxygen in the air.   [1]

14. (a) 9.0 g of aluminium reacts with 35.5 g of chlorine   [1]

$\dfrac{9.0}{27}$ formula masses of aluminium atoms react with $\dfrac{35.5}{35.5}$ formula masses of chlorine atoms.   [1]

0.33 formula masses of aluminium atoms react with 1.0 formula masses of chlorine atoms.   [1]

The formula is given so the marks are awarded for showing the working.

(b) **$2Al(s) + 3Cl_2(g) \rightarrow 2AlCl_3(s)$**   [2]

1 mark is awarded for writing $Cl_2(g)$ and 1 mark for correct balancing.

15. (a) No mass change,   [1]

because none of the products escape.   [1]

(b) Carbon dioxide gas escapes from the reaction mixture.   [1]

Science: On Course for GCSE NEAB Edition Teacher's Book © Stanley Thornes (Publishers) Limited 1998

Module Test 1: Answer grid

QUESTION ONE	1	2	3	4
red cell	☐	☐	☐	■
white cell	☐	■	☐	☐
plasma	☐	☐	■	☐
platelet	■	☐	☐	☐

QUESTION TWO	1	2	3	4
strengthens the bacterium	☐	☐	■	☐
carries genes	☐	☐	☐	■
is where most of the chemical reactions take place	☐	■	☐	☐
controls the entry of substances into the bacterium	■	☐	☐	☐

QUESTION THREE	1	2	3	4
most organs	☐	■	☐	☐
liver	☐	☐	■	☐
lung	☐	☐	☐	■
small intestine	■	☐	☐	☐

QUESTION FOUR	1	2	3	4
red cell	☐	■	☐	☐
white cell	☐	☐	☐	■
plasma	☐	☐	■	☐
platelet	■	☐	☐	☐

QUESTION FIVE	1	2	3	4
alveoli	☐	☐	☐	■
bronchi	☐	■	☐	☐
bronchioles	☐	☐	■	☐
trachea	■	☐	☐	☐

QUESTION SIX	
fruits	☐
margarine	☐
fish	■
pulses	■
root vegetables	☐

QUESTION SEVEN	
brain	☐
digestive	■
kidney	☐
heart	☐
circulation	■

QUESTION EIGHT	A	B	C	D
8.1	☐	☐	☐	■
8.2	■	☐	☐	☐
8.3	☐	☐	☐	■
8.4	☐	☐	☐	■

QUESTION NINE	A	B	C	D
9.1	☐	☐	■	☐
9.2	☐	■	☐	☐
9.3	☐	☐	■	☐
9.4	☐	☐	☐	■

QUESTION TEN	A	B	C	D
10.1	☐	■	☐	☐
10.2	☐	☐	☐	■
10.3	☐	■	☐	☐
10.4	☐	☐	☐	■

Module Test 2: Answer grid

QUESTION ONE	1	2	3	4
chemicals	□	□	□	■
light	□	□	■	□
pressure	□	■	□	□
sound	■	□	□	□

QUESTION TWO	1	2	3	4
bladder	□	□	■	□
liver	□	□	□	■
kidney	□	■	□	□
produces glucagon	■	□	□	□

QUESTION THREE	1	2	3	4
cornea	□	□	□	■
iris	□	□	■	□
pupil	□	■	□	□
sclera	■	□	□	□

QUESTION FOUR	1	2	3	4
chloroplast	□	□	□	■
cell wall	■	□	□	□
vacuole	□	■	□	□
cell membrane	□	□	■	□

QUESTION FIVE	1	2	3	4
phloem	□	□	■	□
root hairs	■	□	□	□
stomata	□	□	□	■
xylem	□	■	□	□

QUESTION SIX	
brain	■
heart	□
kidney	□
liver	■
stomach	□

QUESTION SEVEN	
darkness	□
high humidity	□
low humidity	■
low temperature	□
wind	■

QUESTION EIGHT	A	B	C	D
8.1	■	□	□	□
8.2	□	■	□	□
8.3	■	□	□	□
8.4	□	□	□	■

QUESTION NINE	A	B	C	D
9.1	□	□	■	□
9.2	■	□	□	□
9.3	□	□	■	□
9.4	■	□	□	□

QUESTION TEN	A	B	C	D
10.1	■	□	□	□
10.2	□	□	■	□
10.3	□	□	■	□
10.4	□	■	□	□

Module Test 5: Answer grid

QUESTION ONE	1	2	3	4
hot air	□	□	■	□
iron ore, coke and limestone	■	□	□	□
molten iron	□	□	□	■
waste gases	□	■	□	□

QUESTION TWO	1	2	3	4
copper ion	□	□	□	■
copper sulphate solution	□	□	■	□
positive electrode	■	□	□	□
negative electrode	□	■	□	□

QUESTION THREE	1	2	3	4
calcium	□	■	□	□
copper	□	□	□	■
gold	■	□	□	□
tin	□	□	■	□

QUESTION FOUR	1	2	3	4
chloride	□	□	■	□
hydroxide	■	□	□	□
salt	□	■	□	□
sulphate	□	□	□	■

QUESTION FIVE	1	2	3	4
acidic	□	□	□	■
alkaline	□	■	□	□
metal	■	□	□	□
non-metal	□	□	■	□

QUESTION SIX	
conducts electricity	□
high density	□
liquid at room temperature	■
poisonous vapour	■
silver colour	□

QUESTION SEVEN	
nitric acid	■
potassium nitrate	□
sodium hydroxide	□
zinc chloride	□
zinc oxide	■

QUESTION EIGHT	A	B	C	D
8.1	□	■	□	□
8.2	□	□	□	■
8.3	□	■	□	□
8.4	□	□	□	■

QUESTION NINE	A	B	C	D
9.1	■	□	□	□
9.2	□	■	□	□
9.3	□	■	□	□
9.4	□	□	■	□

QUESTION TEN	A	B	C	D
10.1	□	■	□	□
10.2	□	□	□	■
10.3	□	□	■	□
10.4	□	□	■	□

Science: On Course for GCSE NEAB Edition © Stanley Thornes (Publishers) Limited 1988

Module Test 6: Answer grid

QUESTION ONE	1	2	3	4
granite	■	☐	☐	☐
limestone	☐	■	☐	☐
sandstone	☐	☐	☐	■
schist	☐	☐	■	☐

QUESTION TWO	1	2	3	4
cement	☐	■	☐	☐
concrete	☐	☐	■	☐
glass	■	☐	☐	☐
quicklime	☐	☐	☐	■

QUESTION THREE	1	2	3	4
calcium hydroxide	☐	■	☐	☐
calcium oxide	■	☐	☐	☐
carbon dioxide	☐	☐	■	☐
water	☐	☐	☐	■

QUESTION FOUR	1	2	3	4
crust	☐	☐	■	☐
inner core	■	☐	☐	☐
outer core	☐	■	☐	☐
mantle	☐	☐	☐	■

QUESTION FIVE	1	2	3	4
air	☐	■	☐	☐
fossil	☐	☐	☐	■
hydrocarbons	☐	☐	■	☐
sea creatures	■	☐	☐	☐

QUESTION SIX	
they are on the same plate	☐
they are still moving	■
the Atlantic Ocean between them is getting narrower	☐
similar rocks are present on each	■
the plates will join again in the future	☐

QUESTION SEVEN	
plates are moving together	☐
new rocks are being produced	■
rocks are returning to the mantle	☐
plates are sliding past each other	☐
new magnetic patterns in rocks formed	■

QUESTION EIGHT	A	B	C	D
8.1	☐	☐	☐	■
8.2	☐	☐	■	☐
8.3	☐	☐	■	☐
8.4	☐	☐	■	☐

QUESTION NINE	A	B	C	D
9.1	☐	☐	■	☐
9.2	☐	☐	☐	■
9.3	☐	■	☐	☐
9.4	☐	☐	■	☐

QUESTION TEN	A	B	C	D
10.1	☐	☐	☐	■
10.2	☐	■	☐	☐
10.3	☐	■	☐	☐
10.4	■	☐	☐	☐

Answers

Module Test 9: Answer grid

QUESTION ONE	1	2	3	4
conduction	□	□	■	□
convection	□	■	□	□
radiation	■	□	□	□
insulation	□	□	□	■

QUESTION TWO	1	2	3	4
absorber	□	□	□	■
conductor	□	■	□	□
insulator	□	□	■	□
reflector	■	□	□	□

QUESTION THREE	1	2	3	4
coal	■	□	□	□
hydroelectric	□	□	□	■
nuclear	□	□	■	□
turbines	□	■	□	□

QUESTION FOUR	1	2	3	4
heat	□	□	■	□
light	□	□	□	■
movement	□	■	□	□
sound	■	□	□	□

QUESTION FIVE	1	2	3	4
electricity	■	□	□	□
heat (thermal)	□	□	□	■
sound	□	□	■	□
movement (kinetic)	□	■	□	□

QUESTION SIX	
oil	□
wind	■
wood	■
North Sea gas	□
electricity	□

QUESTION SEVEN	
oil	■
wind	□
wood	□
North Sea gas	■
electricity	□

QUESTION EIGHT	A	B	C	D
8.1	□	□	□	■
8.2	□	■	□	□
8.3	□	□	■	□
8.4	□	□	■	□

QUESTION NINE	A	B	C	D
9.1	■	□	□	□
9.2	■	□	□	□
9.3	□	□	■	□
9.4	□	□	□	■

QUESTION TEN	A	B	C	D
10.1	□	□	□	■
10.2	□	■	□	□
10.3	□	■	□	□
10.4	□	□	□	■

Science: On Course for GCSE NEAB Edition © Stanley Thornes (Publishers) Limited 1988

Module Test 10: Answer grid

QUESTION ONE	1	2	3	4
ammeter	□	■	□	□
diode	□	□	□	■
cell	■	□	□	□
variable resistor	□	□	■	□

QUESTION TWO	A	B	C	D
diode	□	□	■	□
filament lamp	■	□	□	□
polythene rod	□	□	□	■
resistive wire	□	■	□	□

QUESTION THREE	1	2	3	4
live wire	□	□	□	■
neutral wire	■	□	□	□
earth wire	□	■	□	□
fuse	□	□	■	□

QUESTION FOUR	1	2	3	4
current	■	□	□	□
coil	□	□	■	□
core	□	□	□	■
electromagnet	□	■	□	□

QUESTION FIVE	1	2	3	4
generator	□	□	■	□
motor	□	□	□	■
relay	■	□	□	□
transformer	□	■	□	□

QUESTION SIX	
filament lamp	■
polythene rod	□
variable resistor	■
diode	□
open switch	□

QUESTION SEVEN	
limits the amount of current that passes in a circuit	□
melts if the current becomes too great	■
protects the cable from fire hazard	■
is always placed in the neutral connection	□
controls the current that passes in a circuit	□

QUESTION EIGHT	A	B	C	D
8.1	□	□	■	□
8.2	□	□	□	■
8.3	■	□	□	□
8.4	□	□	■	□

QUESTION NINE	A	B	C	D
9.1	■	□	□	□
9.2	□	□	■	□
9.3	■	□	□	□

QUESTION TEN	A	B	C	D
10.1	□	□	□	■
10.2	□	□	■	□
10.3	□	□	□	■
10.4	□	□	■	□